Electroresponsive Molecular and Polymeric Systems

ELECTRORESPONSIVE MOLECULAR AND POLYMERIC SYSTEMS
A Series of Reference Books and Textbooks

Electroresponsive Molecular and Polymeric Systems

VOLUME 1

edited by

Terje A. Skotheim

BROOKHAVEN NATIONAL LABORATORY
UPTON, NEW YORK

 CRC Press
Taylor & Francis Group
Boca Raton London New York

CRC Press is an imprint of the
Taylor & Francis Group, an **informa** business

CRC Press
Taylor & Francis Group
6000 Broken Sound Parkway NW, Suite 300
Boca Raton, FL 33487-2742

© 1988 by Taylor & Francis Group, LLC
CRC Press is an imprint of Taylor & Francis Group, an Informa business

No claim to original U.S. Government works

ISBN 13: 978-0-8247-7968-9 (hbk)

**Visit the Taylor & Francis Web site at
http://www.taylorandfrancis.com**

**and the CRC Press Web site at
http://www.crcpress.com**

Library of Congress Cataloging-in-Publication Data

Electroresponsive molecular and polymeric systems.

Includes index.
1. Polymers and polymerization--Electric properties.
I. Skotheim, Terje A.
QD381.9.E38E395 1988   620.1'9204297   88-10855
ISBN 0-8247-7968-1

About the Series

The science of electroresponsive polymers has become one of the
most expansive and dynamic fields of research in recent years.
Few other areas span to such an extent the whole spectrum of
scientific endeavor, from synthetic organic chemistry to theoretical
physics. The literature is therefore equally scattered, and any
overview is obtained only with great difficulty. There is a need for
a common forum to review the major developments of the field.

This series is devoted to discussing the central topics at a level
that is useful to those actively working in the field as well as to
those seeking entry into a new area. The areas we intend to cover
include recent developments in electronically conducting conjugated
polymers, ionically conducting polymers, redox polymers, piezo- and
pyroelectric polymers, ferromagnetic systems, nonlinear optical
systems, and developments relating to the prospects for achieving
high-temperature superconductivity in organic materials. Other
topics will be covered as they develop. Both molecular and polymeric
systems will be covered.

Developments described in this series represent a dramatic expansion of the scope of polymer science, from being a field dealing
primarily with structural and insulating materials to one that includes materials with a potentially endless variety of electrical and
optical properties. With this leap we have truly come to the threshold of a new era, not only in polymer science, but in materials science

itself. The prospect of employing the skills of synthetic chemists to tailor materials with electrical and optical properties as precise and specific as those of biological systems opens vistas we have not seen before in materials science. The extraordinary technological potential of these materials will be highlighted in a number of forthcoming chapters on applications.

The purpose of this series is to provide detailed, topical reviews of current developments in the various subfields of electroresponsive polymer science. The intent is to highlight recent developments and provide a broader perspective on the field as a whole. By bringing together these separate fields of polymer research, it is hoped that the series will provide a focal point to stimulate future developments.

Preface

It is now a little more than ten years since the beginning of the modern era of electroresponsive polymer science. The pace of its development has been extraordinarily fast, a fact that is testimony to the creative dynamics of a highly interdisciplinary field. The hurdles that must be surmounted to achieve a detailed understanding of the structure and transport mechanisms on the one hand and the synthesis of commercially useful materials on the other have fallen much faster than was thought possible only a few years ago.

Some significant breakthroughs have taken place during the last few years. From a technological point of view, the synthesis of processable and stable electronically conducting polymers certainly represents a major advance toward large-scale commercial applications. Perhaps the most spectacular development has been the synthesis by Herbert Naarmann and co-workers of a more oriented, defect-free form of polyacetylene, with exceptionally high conductivity and enhanced environmental stability. This is highlighted in the chapter by Naarmann and Theophilou, who give a general overview of recent work in the synthesis of new materials. The results with Naarmann polyacetylene have raised questions about the adequacy of current models of the conductivity mechanisms. Some recent theoretical developments are treated in the chapter by Soos and Hayden.

Polyconjugated systems have also exhibited exceptional nonlinear optical properties due to the high polarizability of the extended π-electron system. The possibilities for applications in the area of nonlinear optics do indeed look intriguing, and it is a topic we will develop in future volumes of this series. Piezo- and pyroelectric polymers are already incorporated in commercial devices, such as high-quality microphones. A review of their status is given in the chapter by Marcus.

The ion-conducting polymers represent another area where major developments in theory, synthesis, and technological applications have taken place during the last few years. Gauthier, Armand, and Muller summarize the present understanding of structure and transport properties in this class of materials. They also give an account of the state of the art of the development of thin-film rechargeable lithium batteries incorporating an ion-conducting polymer as thin film solid electrolyte. These batteries have very high energy storage capacities and cyclability and may eventually represent a large-scale commercial introduction of conducting polymers.

The extraordinary variety of possible structures and properties of electroresponsive polymeric systems is also highlighted in the chapter by Abruña, who discusses redox polymer systems and their potential analytical and catalytic applications.

The chapters in this first volume of the series are intended to give a representative overview of the state of the art of several of the subfields of electroresponsive polymer science. In the future, entire volumes may be dedicated to specific topics that warrant a more complete and detailed coverage.

The field of electroresponsive polymer science is clearly maturing, and we are now witnessing a rapidly growing worldwide effort in both university and industrial laboratories to capitalize on the emergence of electroresponsive polymer research as a major field of materials science. It is hoped that this series will provide a sufficiently broad perspective to stimulate developments in new directions and to bring researchers up to date on the field as a whole.

On behalf of the editorial board, I wish to express my gratitude to the authors for their valuable contributions and to the staff of Marcel Dekker for their cooperation in making this series a reality.

Terje A. Skotheim

Contents

Contributors

Héctor D. Abruña   Department of Chemistry, Baker Laboratory,
   Cornell University, Ithaca, New York

Michel Armand   Laboratoire d'Ionique et d'Electrochimie du Solide,
   LIES ENSEEG Domaine Universitaire, St-Martin d'Hères, France

Michel Gauthier   Institut de Recherche d'Hydro-Québec (IREQ),
   Varennes, Quebec, Canada

G. W. Hayden   Department of Chemistry, Princeton University,
   Princeton, New Jersey

Michael A. Marcus   Analytical Technology Division, Eastman Kodak
   Company, Rochester, New York

Daniel Muller   Groupement de Recherche de Lacq, Groupe
   Elf-Aquitaine, Lacq, France

Herbert Naarmann   BASF Plastics Research Laboratory, Ludwigshafen,
   West Germany

Z. G. Soos   Department of Chemistry, Princeton University,
    Princeton, New Jersey

N. Theophilou   BASF Plastics Research Laboratory, Ludwigshafen,
    West Germany

Electroresponsive Molecular and Polymeric Systems

1

Synthesis of New Electronicaly Conducting Polymers

HERBERT NAARMANN and N. THEOPHILOU / BASF Plastics
Research Laboratory, Ludwigshafen, West Germany

Who can find a wise and original idea which hasn't been thought
before?

Goethe, *Faust*

I. INTRODUCTION

Conductive polymeric materials widen the horizon in science and
technology, and it is not only chemists to whom they present a challenge. This chapter is an attempt to demonstrate the variety of
scientific thought required to explore the various stimulating areas
of organic chemistry involved. The scope is far too wide to be
covered by one contribution.

There has been an avalanche of publications on the subject. Those
written in English up to 1984 have been reviewed by W. J. Feast
[1], and those in German up to 1982 have been the subject of other
reviews [2,3]. In this contribution, it is taken for granted that the
results described in these reviews are general knowledge, and we
therefore wish to concentrate on more recent developments. They
are presented from the viewpoint of a chemisty who had already prepared conductive polymers in the 1960s and who still has not lost
his interest in the correlation between structure and properties.

II. BASIC SYSTEMS AND PRINCIPLES

Many routes are available for the synthesis of the polymers' conductive backbones. They include Wittig, Horner, and Grignard reactions,
polycondensation processes, and metal-catalyzed polymerization techniques. Oxidative coupling with oxidizing Lewis acid catalysts
generally leads to polymers with aromatic or heterocyclic building
blocks.

The Ziegler-Natta process has been widely adopted for the production of film-forming polyacetylenes, but it is only one of the many
feasible reactions in which the catalysts are combinations of reducing
agents and transition metals. The polymers thus obtained are made
conductive by doping with n-type materials by reduction and with
p-type materials by oxidation.

Direct polymerization of acetylene or aromatics in which the dopant
is incorporated or which proceeds via anodic oxidation (e.g., of
pyrrole) leads to highly conductive materials.

No plastics or other organic polymers display the beneficial combination of properties peculiar to metals, i.e., (a) great mechanical
strength, (b) ductility, and (c) thermal and electrical conductivity.
Conductive metals have a high specific gravity, whereas organic
polymers are lighter and are typical dielectrics. It is the combination

of low weight and high conductivity that makes the new class of synthetic materials particularly interesting from an economic point of view and assures them of a wide field of application.

In the last two years, astonishing progress has been made in attaining greater conductivity and stability. If metals are compared with polyacetylene on a basis of the conductivity/weight ratio, it can be seen that the polymers have already reached the same level as that of many metals (Fig. 1).

The aim of the syntheses is to produce conjugated systems with the maximum possible number of mobile π-electrons.

The underlying structural principles (I–III) are shown in the following scheme:

Polyacetylene: trans 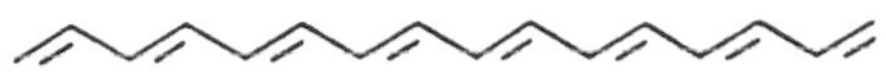

cis 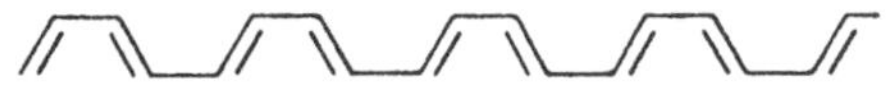 I

Polyphenylene: 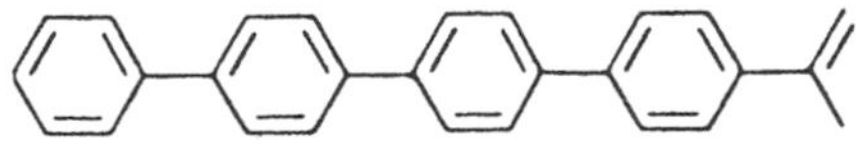 II

Polypyrrole: 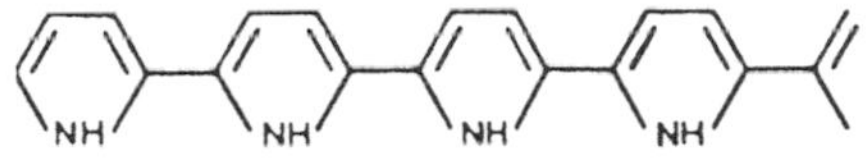 III

Polymers with π-electron conjugation

The first step in all the syntheses is to produce the basic structure with extended unbroken electron conjugation. In a subsequent step, the basic structure is converted into the conductive polymer by oxidation or reduction. Exceptions to these scheme are electrochemical oxidation (e.g., of pyrrole) and gas-phase polymerization (e.g., with arsenic (V) fluoride). These reactions lead directly to conductive polymers in a single-stage process, that is, without isolation of the basic structure.

The first structural principle (I) represents the polyenes; the second (II), the polyaromatics including polycondensation types; and the third (III), the polyheterocycles. The common feature of all structures is the presence of an extended π-conjugation.

Exceptions to these structures are hetero atoms, for example —S—, or hetero groups, for example

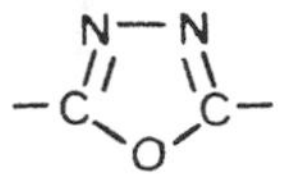

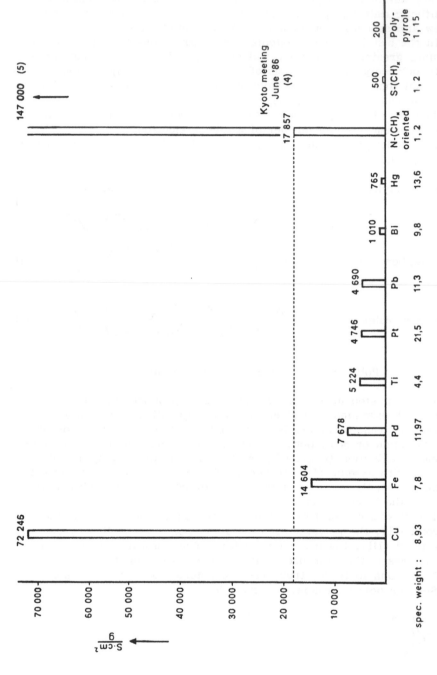

Fig. 1　Comparison of the electrical conductivity (by weight) of different metals and some electrically conducting polymers.

bridged aromatics, and also polysilanes, whose conductivity can be ascribed to increased polarization or σ-electrons.

A review of recent trends has shown that scientific progress and practical application depend on the following:

1. Reproducible preparation of specimens of extensive area
2. Determination of the conditions for synthesis and the laws relating them to properties

For this reason, the following sections have been compiled in all cases from the aspect of novel products or improvements in production processes.

III. IMPROVEMENTS ON EXISTING PROCEDURES

A. New Syntheses: Conditions and Methods

(CH)$_x$-Polymerization in Silicone Oil—a Room Temperature Process. A decisive step towards the production of polyacetylene was the determination of the direct relationship between conductivity and crystallinity and the indirect relationship to the number of sp^3 orbitals [6].

If the polymerization conditions are modified (e.g., by silicone oil), a new polyacetylene type, N—(CH$_x$), can be obtained at room temperature. Its quality is at least as good as that of the standard polyacetylene-type S—(CH)$_x$ prepared by the Shirakawa technique at −78°C. The results are presented in Table 1 [7].

Aging of the standard catalyst is responsible for another, surprising improvement in the properties of the (CH)$_x$. The reduction in the number of sp^3 orbitals, i.e., the production of a system free from defects, is a great advantage [4].

The work carried out by Tanaka et al. [11] was in the same direction, that is, the avoidance of impurities. Extremely pure (CH)$_x$ with remarkably low initial conductivities (2 × 10^{-14} S/cm) were obtained.

(CH)$_x$: Orientation by Mechanical Stretching. Orientation of thin (CH$_x$) films leads to surprisingly high conductivity values as shown in Fig. 1.

Figure 2 shows the differences in morphology between S(CH)$_x$ and oriented N(CH)$_x$.

(CH)$_x$: Highly Conductive and Transparent. Transparent (CH)$_x$ film with conductivities higher than 5000 S/cm can be prepared by the method mentioned above [5]. The polyacetylene is produced on a plastic film and stretched along with the supporting material. Afterwards, it can be peeled off to yield a self-supporting film which can be doped under standard conditions.

Table 1 Properties of the Different $(CH)_x$ Types

	Crystallinity[a] [%]	Conductivity[b] [S/cm]		sp^3 content[c] [rel. %]	Surface area[d] [m^2/g]	Configuration[e] cis content [%]
		Undoped	Doped with I$_2$			
S—$(CH)_x$	70	10^{-6}	200	4	300	50
N—$(CH)_x$	65	10^{-8}	2000	0	100	80

[a]Phillips defractometer Cu K$_\alpha$ radiation.
[b]Four probe measurement.
[c]Determined by 13C-NMR spectroscopy.
[d]BET-method.
[e]Determined by IR spectrometry.

An almost perfect alignment of the stretched $(CH)_x$ envelopes has been described by G. Lugli et al. [42].

Highly Conductive $(CH)_x$ by Polymerization in an Oriented Matrix. This method, adopted by Shirakawa [8] and Hitachi KK [14], was developed by Aldissi [9]. It offers a means of polymerizing $HC \equiv CH$ in an oriented matrix of nematic liquid crystals. The effect of the orientation depends on the field strength and the type of liquid crystals.

Shirakawa attained values of 12,000 S/cm [8a].

A Japanese patent held by Polyplastics KK [13] describes the $CH \equiv CH$ polymerization in a matrix consisting of a resin that displayed anisotropy in the melt phase.

New Synthesis Route for the Preparation of (CH_x).   Costello and McCarthy reported that the reduction of polytetrafluoroethylene by

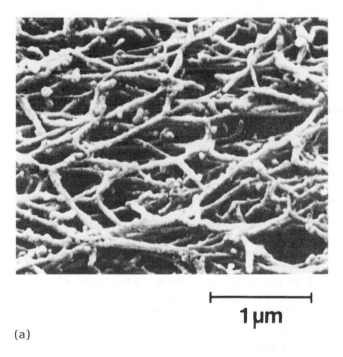

1 µm

(a)

Fig. 2 Electromicrographs of standard polyacetylene, S—$(CH)_x$ and new polyacetylene, N—$(CH)_x$. (a) S—$(CH)_x$: conductivity (iodine-doped): 200 S/cm. (b) N—$(CH)_x$: conductivity (iodine-doped): 10,000 S/cm. (Ref. 5.)

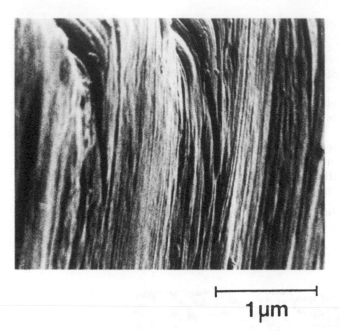

1 µm

(b)

Fig. 2   (Continued)

the benzoin dianion in dimethylsulfoxide gives rise to a gold-coloured,
air-sensitive film [13a].   The conductivity obtained after iodine oxidation was 36 S/cm.   The probable reaction is shown in Figure 3.

Continuous Preparation of Polypyrrole.   Polypyrrole and other
polyheterocycles are more stable than polyacetylene and can be produced quite easily by electrochemical techniques [10a,10b] (Fig. 4).
The continuous process yields homogeneous material in the form of
self-supporting film of 20 µm to 500 µm gauge.   Properties of the
material depend on preparation conditions, e.g., current density,
pH, the electrolyte, counterions, and monomers [10].

Orientation of Polypyrroles.   The work carried out by the Agency
of Industry Science Technology has attracted great attention.   Degrees of orientation of 90% and conductivities of 1005 S/cm are attained if polypyrrole film produced at −50°C is subsequently
mechanically stretched at room temperature and briefly exposed to
150°C [12].

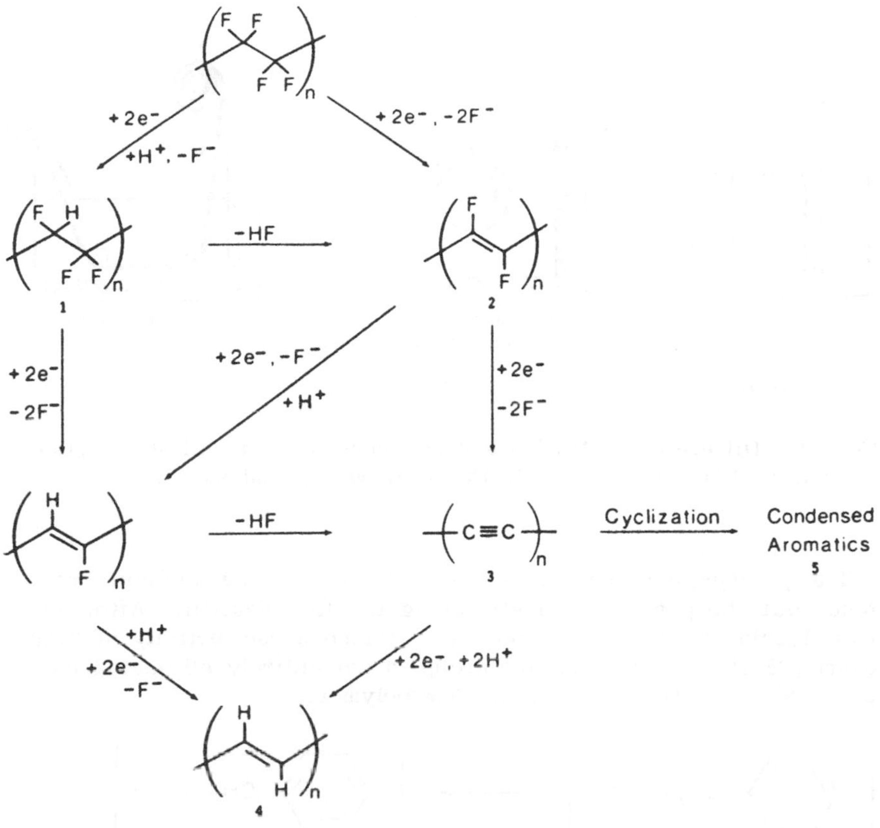

Fig. 3   Possible scheme by which polytetrafluoroethylene is reduced by benzoin and potassium tert-butoxide in dimethylsulfoxide solution. (Ref. 13a.)

Other Syntheses.   Highly transparent and conductive polypyrrole/ poly(vinyl alcohol) composite film can be prepared by gas-phase polymerization, in which gaseous pyrrole is allowed to act on poly-(vinyl alcohol) film containing iron(III) chloride.   The conductivity correlates positively with the iron(III) chloride fraction; the transparency, negatively [49].

In the Polaroid process [70] for the production of processable conductive pyrroles (as well as thiophenes and furans), preference is given to counterion polymers in polymeric latex particles.   Polymerization and incorporation of the five-ring heterocycles in the special latex are effected electrochemically.

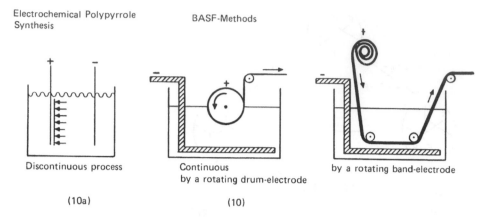

Electrochemical Polypyrrole
Synthesis

BASF-Methods

Discontinuous process

Continuous
by a rotating drum-electrode

by a rotating band-electrode

(10a)

(10)

Fig. 4   Different electrochemical polymerization techniques.   [(a)
from Ref. 10a; (b) from Ref. 10, used with permission.]

The poly(p-phenylene vinylene) synthesis via the sulfonium route
holds out the possibility of stretching the fibrillization.   After arsenic-
(V) fluoride oxidation, the polymer attained a conductivity of 2780
S/cm [25,26].   The sulfonium group is not entirely eliminated by
pyrolysis from the resulting random polymer.

$$\left[-\left< \right> -CH_2-CH- \right] \quad \longrightarrow \quad \left[-\left< \right> -CH=CH- \right]$$
$$\underset{\underset{CH_3 \quad CH_3}{|}}{S^+}$$

B.   New Catalyst Systems

Two noteworthy trends in catalysts for $(CH)_x$ syntheses are emerging.
One is defined annealing of the catalysts.   Remarkable changes in
properties have been achieved, and the stability of the polyacetylene
has been improved, i.e., its susceptibility to oxygen has been reduced [15,16].   The other is the use of catalyst combinations, namely,
salts of transition metals in Groups IV–VIII in the periodic system
and reducing agents, preferably lithium alkyls.   Some examples follow:

Tungsten(VI) chloride/n-butyllithium [17]
$Ti(O\text{-}butyl)_4$/n-butyllithium [18]
$Ti(O\text{-}butyl)_4$/ethylmagnesium bromide [19]
$Ti(benzyl)_4$/n-butyllithium [20].

$FeCl_3$ ($HfCl_4$, $TaCl_5$, $NbCl_5$, $MoCl_5$, $ZrCl_4$, VCl_3, $CrCl_3$)/
n-butyllithium [20a].

These catalyst systems are normally used in reaction media consisting
of toluene or silicone oil.

An extremely interesting synthesis is known as the ARA method.
An aged mixture of Ti $(OC_4H_9)_4$/Al$(C_2H_5)_3$ in silicone oil, mixed
with various quantities of n-butyllithium and allowed to react with
acetylene, yields polyacetylene that is highly regular, compact and
crystalline in well-defined parallel planes [21]. This film can be
stretched by up to 400% ($1/1_0 = 5$) and, after doping, has a volume
conductivity of 100,000 S/cm. Due to its specific weight of about
1.2 (see Fig. 1), its weight conductivity is in the same range,
namely, 100,000 S·cm^2/g.

A variant of the Luttinger-Greene $(CH)_x$ synthesis uses the transition metal alkoxyl Me(alkali$(OR)_x$). For instance, nickel(II) bromide
reacts with methoxylithium in tetrahydrofuran. The catalyst thus
obtained gives rise to trans-$(CH)_x$ with a yield of about 88% [24].

High-Pressure Solid-State Polymerization. 1,4-*Bis*(p-toluene-
sulfonyloxymethyl)1,3-butadiyne is polymerized at pressures of up
to 8.3 GPa and subsequently electrochemically oxidized with silver
chlorate. The conductivity thus obtained is about 2 S/cm [89].

Tungsten carbonyl complexes are efficient catalysts for synthesizing
$(CH)_x$ by insertion polymerization [22]. Soluble polymers are thus
obtained.

$R = C_5H_{11}$ n = 1,2,3...

Bis(benzene)chromium-(O) is an effective initiator for polymerizing
tetrafluoroalkynonitrile. Conductivities of 10^{-8} S/cm are thus attained [88].

A photochemical method for the production of polyparaphenylene
consists of decomposing p-diiodobenzene by irradiation under a low-
pressure mercury-vapor lamp. Specimens of up to 0.27 μm are thus
obtained [23].

Polymerization reactions triggered by arsenic(V) fluoride now include condensed phenols and mercaptobenzene [45] and pyrrole [68].
The first step in this direction was the single-stage acetylene polymerization with simultaneous doping [47]. Another publication [50]
describes the single-stage process for the production of conductive

polyaromatics. Frommer reported on combinations of arsenic(III) fluoride as the solvent and arsenic(V) fluoride as the dopant [58].

The nickel- and palladium-catalyzed dehalogenation of dihaloaromatic compounds with zinc is a new route to poly(2,5-thienylenes) and poly-(1,4-phenylene) [46].

An interesting variant for polyphenylenes is electrochemical oxidation with the aid of copper(II) chloride/lithium arsenofluoride. The process gives rise to flexible films [48].

Plasma polymerization represents an interesting means of producing conductive polymers. Acetonitrile, for example, is polymerized under reduced pressure and at 100 W in a plasma generator. The polymer forms as a film on a prepared polyurethane. After treatment with copper sulfate and sodium thiosulfate, its conductivity is of the order of 10^{-3} S/cm.

It has been stated that pyrrole and compounds containing cyanide groups can also be plasma-polymerized [62,87].

C. New Reactants

This section deals with significant changes in structure and properties of conductive polymers that can be achieved by varying the reactants.

Polyacetylenes

Soluble Block Copolymers of Acetylene. Polyolefin blocks, produced by anionic polymerization, are subsequently treated with tetrabutoxytitanium and acetylene gas to yield soluble block polymers with conductivities of 1–10 S/cm [27]. Reports on analogous work have been published by Destri et al. [28], Schué et al. [29a], Aldissi et al. [29] and Galvin et al. [30], who produced polybutadiene and polystyrene block copolymers.

$(CH)_x$-Polymerization in the Presence of Polymer Solutions. Instead of in toulene, the conventional Shirakawa techniques are carried out in the presence of polymer solutions, e.g., polyisobutenes or polyethylene. The polymer films thus obtained contain mass fractions of up to 20% of polymers that are stretchable but, after doping, are no more stable towards oxygen than polyacetylene [31].

Polymerization of Acetylene in the Presence of Polystyrene or Polyoxyethylene. This process has been described by ELF Aquitaine [32]. Conductivities after iodine oxidation are of the order of 800 S/cm.

Poly(acetylene-10-ethylene) yields polymers with conjugated segments of up to ten double bonds via the isomerization of poly(1,3-butadiene) with the aid of potassium *tert*-butylate [41].

Polyheterocycles

Substantial progress has been achieved in the electrochemical polymerization of heterocycles.   There has been considerable work on
influencing polymer properties with counterions (Fig. 5).

Large counterions, such as ligninsulfonate, improve polymer
stability.   Counterions containing quinone groups influence redox
conditions.   Toyota researchers described porphyrin trisulfonate as
a counterion in water [69].   Other compounds that attract interest
in this connection are: pyrrolesulfonic acid and, by analogy, thio-
phenesulfonic acid [36,36a] and other sulfonic acids of pyrrole and
thiophene oligomers [35].

$$X = (CH_2)_2 , (CH_2)_4$$

The monomer and the counterion are both present in these molecules.
Thus a decidedly higher charge concentration has been observed in
the polymerization of pyrrolesulfonic acid by itself or in the presence
of pyrrole.

The counterion alternates with the pyrrole in the pyrrole-squaric
acid polymer [37].

Work aimed at improving film properties, e.g., in the electrochemical oxidation of pyrrole or aniline films, has been carried out
with fatty amines and heterocyclic amines [39].

In analogy to anionic polymerization, e.g., with lithium-initiated
starters, the principle of grafting has also been realized for p-polypyr-
role films.   Cationic-polymerizable monomers, such as vinyl ethers, form
graft polymers, whose thickness depends on reaction conditions [96].

A potential variant for widening the scope of $(CH)_x$ consists of coating the unstable polyacetylene by electrochemical polymerization of
pyrrole or thiophene at the $(CH)_x$ electrode [43,38].   The first
publication on this subject appeared in 1984 [40].

Modifying polypyrrole films with poly(platinum/rubidium bipyridine)
complex is a means of producing bilayer electrodes [44].

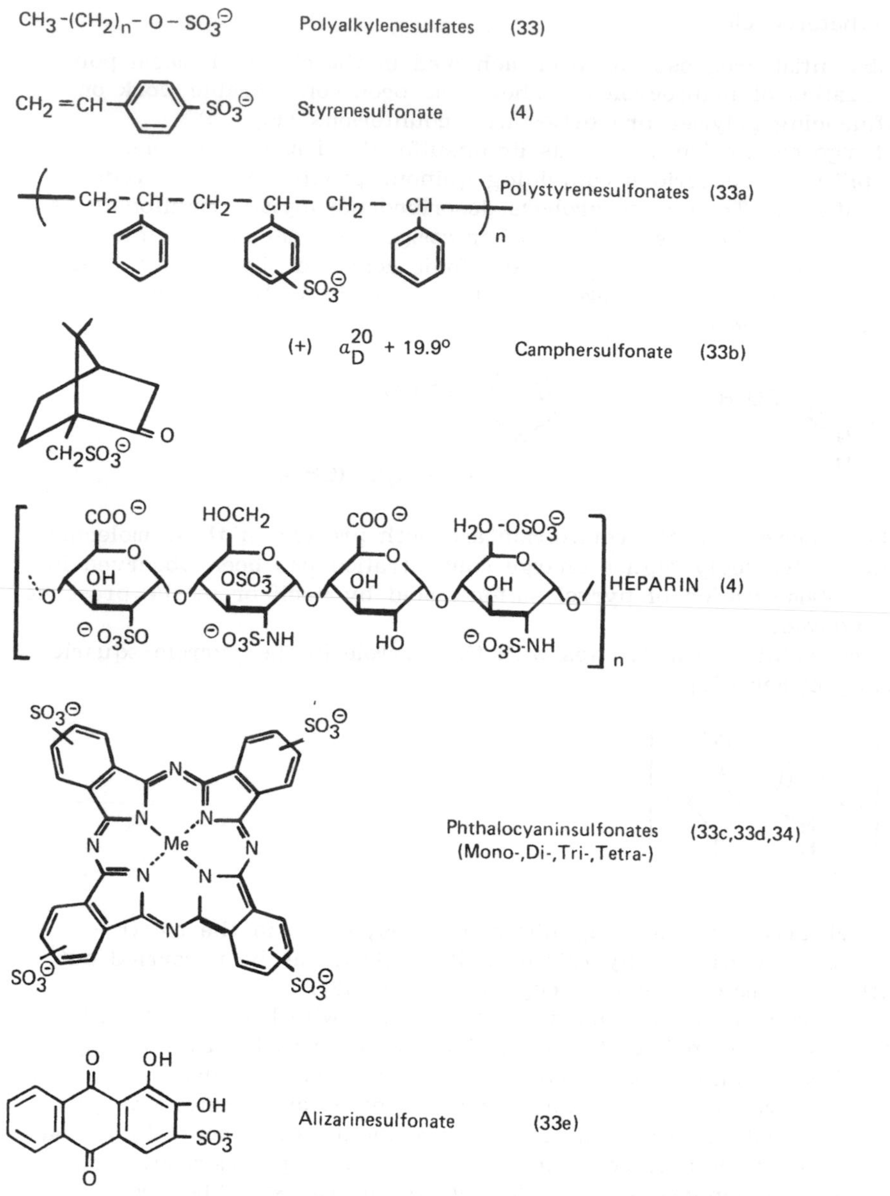

$CH_3-(CH_2)_n-O-SO_3^{\ominus}$ Polyalkylenesulfates   (33)

$CH_2=CH-\langle\!\!\!\bigcirc\!\!\!\rangle-SO_3^{\ominus}$ Styrenesulfonate   (4)

Polystyrenesulfonates   (33a)

(+) a_D^{20} + 19.9°   Camphersulfonate   (33b)

HEPARIN   (4)

Phthalocyaninsulfonates   (33c,33d,34)
(Mono-,Di-,Tri-,Tetra-)

Alizarinesulfonate   (33e)

Fig. 5   Conductive salts that can be incorporated as counterions during the synthesis of polypyrrole.

IV. HOMOLOGUES AND DERIVATIVES OF
TRADITIONAL MONOMERS

Polyacetylenes

A large number of substituted acetylenes have been polymerized by a wide variety of methods (Table 2 lists substituents that have been used.)

Heterocycles—Substituted Pyrroles

The work carried out in this field concerns the investigation of the effects exerted by substituents on polymerization and on such properties as charge fixation and stability etc. In addition, most of the studies are devoted to increasing the usefulness of these increasing polymers by improving their solubility or processability. The results obtained are summarized in Table 3.

Pyrrole-Styrene Graft Copolymers

Poly(4-chloromethylstyrene) reacts electrochemically with pyrrole to yield this polymer:

The films have a conductivity of 0.4 S/cm and can be processed by the techniques adopted for thermoplastics [67].

Thiophene and Selenium/Tellurium Analogues

Substituted Thiophene.   As the length of the alkyl side chains substituted at the sulfur atom of thiophene increases (up to 3-eicosylthiophene), the electrochemically produced polymers become more soluble and less conductive (down to 10^{-1} S/cm) [72].   3-Methylthiophene is an exception, yielding polymers with conductivities of

Table 2   Substituted Acetylenes

Substituent	References
$R_4-Si-R_3-R_4$ R_3 and R_4 = H; a one- to eight-C monovalent hydrocarbon, or a Si-containing monovalent organic group	51, 59
$-C\equiv C-R$ R = alkyl, phenyl, benzyl, naphthyl, or carbazole	52
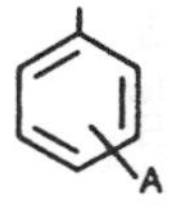 (phenyl with A) A = alkyl, aryl, alkoxy, aryloxy, nitro, or CN	53
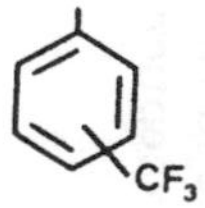 (phenyl with CF_3)	54
$-X$ X = phenyl, thienyl, pyridinyl, or furanyl	55
$-Y$ Y = adamantyl, allyl, aryl, aralkyl, alkoxy, or halogen	56

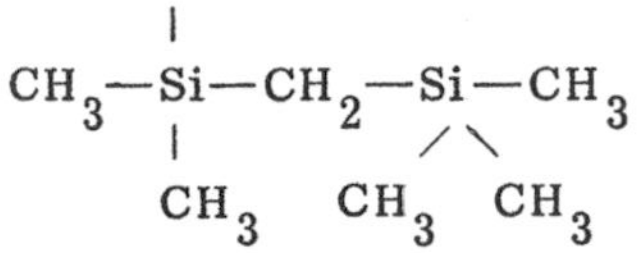

57

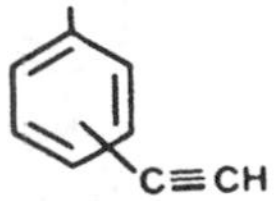

60

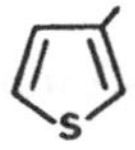

71

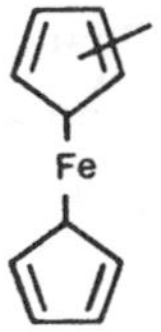

86

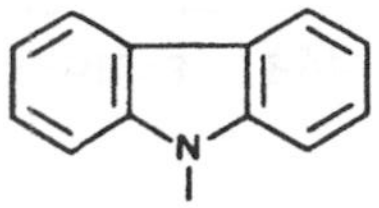

100

"See footnote ^{a}"

^{a} 1,6-bis(9-carbazolyl)-2,4-hexadiyne is reacted in analogy to the carbazolyl derivative. Both polymerizations are carried out in molten iodine with simultaneous doping to conductivies of up to 10^{-3} S/cm. It is known [61] that substituted acetylenes can be polymerized by various methods: metal-catalyzed, cationically, by radicals, and anionically only acetylenes with electron-withdrawing substituents.

Table 3   Pyrrole Variants

Variant	Comment	Reference
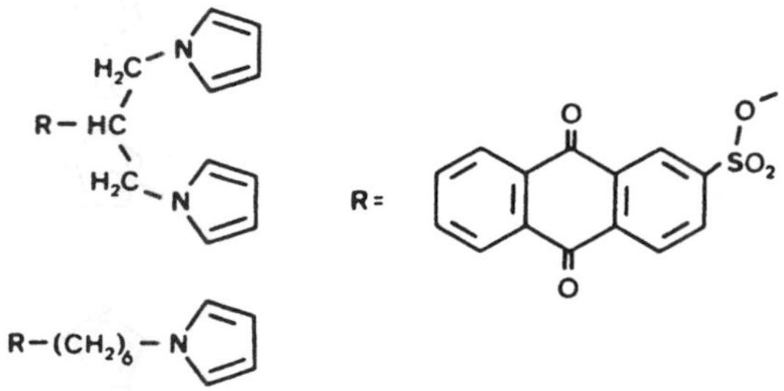	The aim is to prepare polymers containing defined redox partners and to allow the quinone structure to participate in the electrochemical charging and discharging mechanisms	63
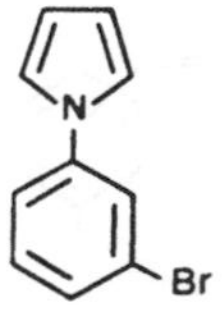	The N-substituted derivative is de-activated. During electrochemical copolymerization with pyrrole, it is restructured and incorporated in a ratio of 9:1 in the polypyrrole.	64

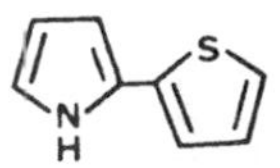

2,2'-thienylpyrrole can be polymerized electrochemically. The conductivity depends on the electrolyte and lies between 0.08 and 3.3 S/cm.

98

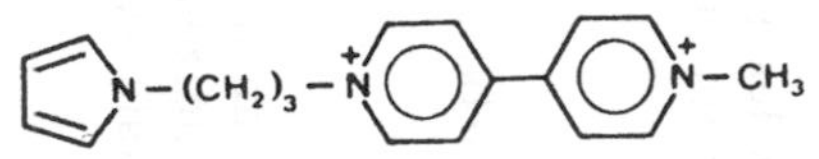

This variant yields polypyrrole films that are transparent and display electrochromic effects.

65

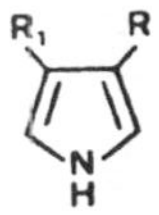

The advantage of the powder polymer, which has a conductivity of 10^{-5} to 10^2 S/cm, is its solubility in aprotic, aromatic, and hetero-aromatic solvents

66

R = hydroxyphenyl

R_1 = alkyl, etc.

150 S/cm. Great scientific value is attached to the observation that polymers of this compound are helical [73], particularly since studies on other polymers have indicated that firm opinions on structures may have to be revised [4].

Selenophene and tellurophene are polymerized electrochemically analogously to thiophene. The polymers are of interest in photo-electronics and as electrophotographic-sensitive materials [74,75,76].

Polycondensed Thiophene-Based Polymers

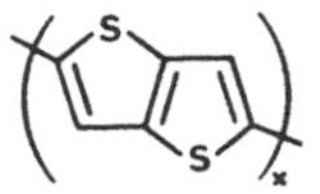

PTT
10^{-6} S/cm

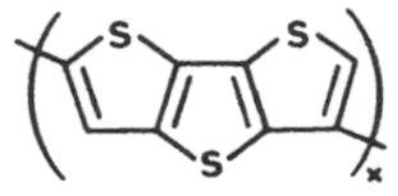

PDTT
10^{-3} S/cm

Polythieno(3,2b)-thiophene (PTT) and Polydithieno(3,2-b; 2'-3'-d)-thiophene (PDDT) [83,85] are obtained by electrochemical oxidation.

Aminoaromatic Structures

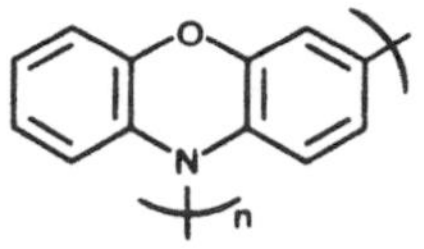

Polyoxazine

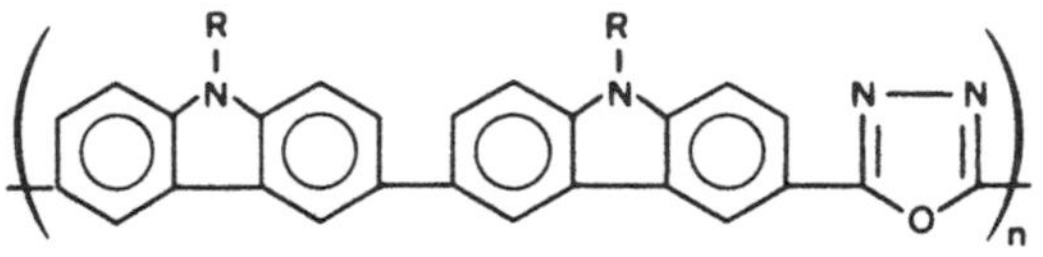

Polybicarbazole-oxadizoles

Polyoxazine [77] and Polybicarbazole-oxadiazoles [78] are processable polymers that can be reversibly oxidized or reduced. They mainly attract attention as photovoltaic systems.

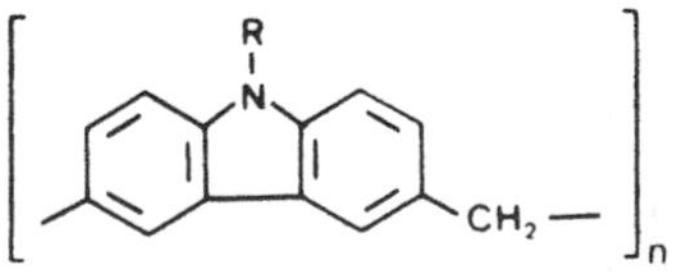

Polyalkylcarbazoles

Polyalkylcarbazoles [79] and polyaminopyrimidines [80,99] are also processable polymers that can be reversibly oxidized or reduced. They merit interest as electrodes for photovoltaic systems.

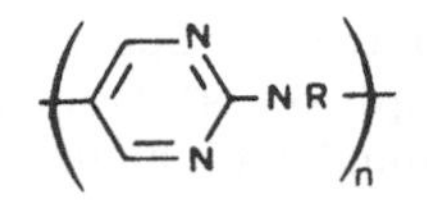

Polyaminopyrimidines

Poly(3,3'-N-substituted carbazolyl) compounds are complexed with iodine and attain conductivities of up to 5 S/cm [101].

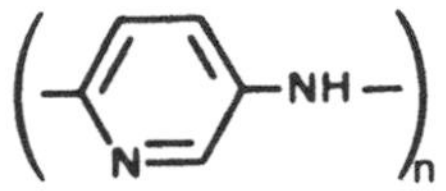

Polyaminopyridines

Polyaminopyridines [111] are prepared by oxidative polymerization. They are stable in air and have conductivities of 10 S/cm.

Cross-linkage With Sulfur

Polythiocarbazoles are produced in reactions between sulfur and carbazole.   They are insoluble n-type semiconductors [84].

Work on the polyaniline complex has been extended to include the polymerization of substituted aminoaromatics (e.g., N-methylaniline, N-ethylaniline, 1-pyreneamine, and o-phenylenediamine), presumably to avoid the formation of soluble benzidine, which is obtained in the aniline oxidation.   The conductivities of the polymers, which can be reversibly charged and discharged, lie between 10^{-3} and 10^{-4} S/cm. As a comparison, the figure for analogously produced polyanilines is 10^{-11} S/cm [81].   Angelopoulos has described new metallic derivatives of polyaniline, prepared by N- and ring substitution [81a].

Copolymerization of aminoaromatics with pyrroles [82] gives rise to films that contain aromatic structural units and feature good electro-chemical reversibility and outstanding stability on exposure to air.

V.   NEW POLYMER STRUCTURES

A.   Conjugated Structures

Shirakawa et al. [90] synthesized carbyne, a new type of polymer, by dehydrochlorination of stereospecifically chlorinated polyacetylene formed by prolonged iron(III) chloride doping.   Iodine oxidation gives rise to conductivities of 10^{-1} S/cm.   The starting value was 10^{-9} S/cm.

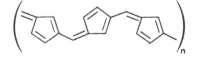

Poly(1,3-cyclopentadienylene methylene)

Poly(1,3-cyclopentadienylene methylene) is obtained from propargyl halides [91]. Polymerization is initiated by metal complexes at temperatures higher than 100°C. The products are cross-linked and have conductivities of 10^{-1} S/cm after they have been n- or p-doped.

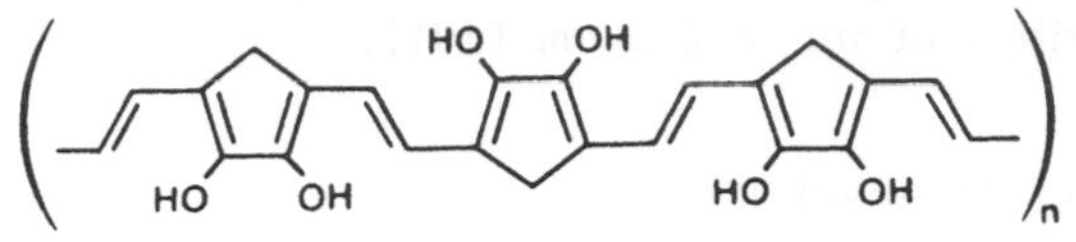

Poly(4,5-bishydroxy-1,3-
Cyclopentadienylene *trans*-vinylene)

Poly(4,5-bishydroxy-1,3-cyclopentadienylene *trans*-vinylene) and structures of this type are synthesized according to the following scheme [92]:

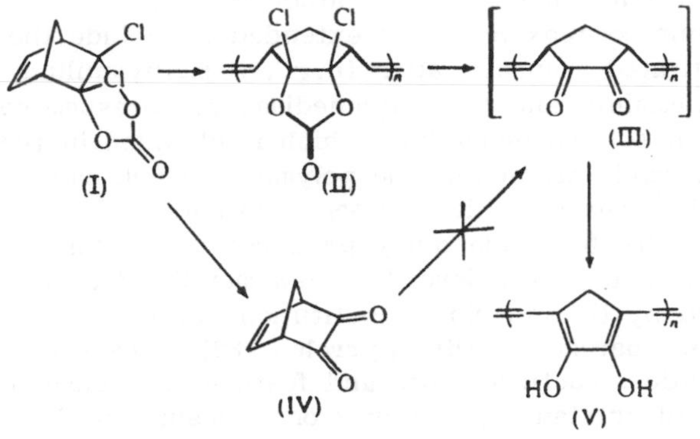

The maximum conductivities are of the order of 10^{-7} S/cm.

Poly(2,5-furanyl) vinylene can be produced from 5-chloromethyl-2-furancarbaldehyde [112]. It can be processed in the form of solutions, e.g., transferred to film, and can be reversibly charged and discharged. Conductivities of up to 7 S/cm can be achieved by oxidation with iodine.

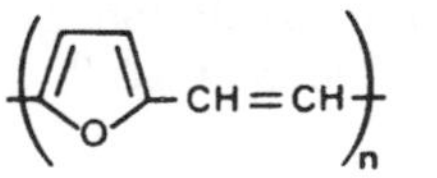

Polymers with phenalenone units [93] are very stable redox compounds that contain chelated metal ions. The conductivity is of the order of 10^{-1} S/cm.

Schopov et al. produced polydiphenylacetylene from 1,2,3-triphenyl prop-2-en-1-one (TPEO) [94].

$$CH = C - C = O \xrightarrow[-C_5H_5-CHO]{WCl_6} \left[\begin{matrix} = C - C = \\ | \quad | \\ C_6H_5 \; C_6H_5 \end{matrix} \right]_n \text{ or } \left[\begin{matrix} - C = C - \\ | \quad | \\ C_6H_5 \; C_6H_5 \end{matrix} \right]_n$$

The polymers are soluble and stable to heat.

Bräunling et al. [95] have described a new route to polyarene-methides:

The products are films or powders with conductivities less than 10 S/cm.

Polyene-fulvenes [4] are produced by cationic polymerization [97] of fulvenes followed by dehydrogenation.

monomer   isolated   conjugated
               double bonds   double bonds

The products are films which have conductivities of up to 100 S/cm after iodine oxidation.

The polymethacyclophanes are a new class of polyconjugated systems [102, 103]:

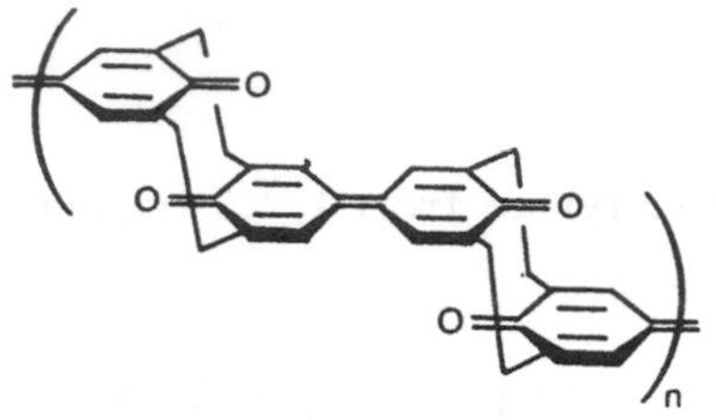

Q-PMCP 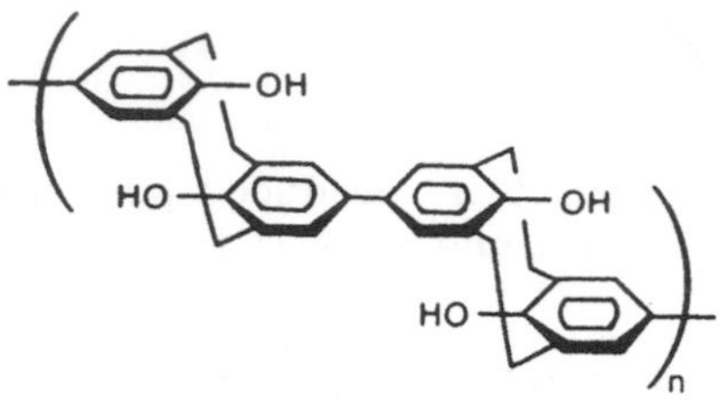   P-CMCP

The longitudinal π-electron overlap plays an important role in conductive Charge-Transfer complexes. Q-PMCP has a [2,2]-metacyclophane unit, a strong transannular π-electron interaction, and a quinone linkage. P-CMCP is the hydroquinonic type. Both products have high crystallinity and reach conductivities of 10^{-1} S/cm with sulfuric acid or iodine doping.

Pioneer work has been done by Staab [104], who developed quinone units containing porphine-ring systems.

B. Poly(β-diketones)

Diketene can be polymerized to yield a polydiketone (PK); the result of silylation is PSiK. This treatment fixes the conjugated double bond and increases the solubility in organic solvents. The conductivity of PK and PSiK is of the order of 10^{-5} S/cm [108].

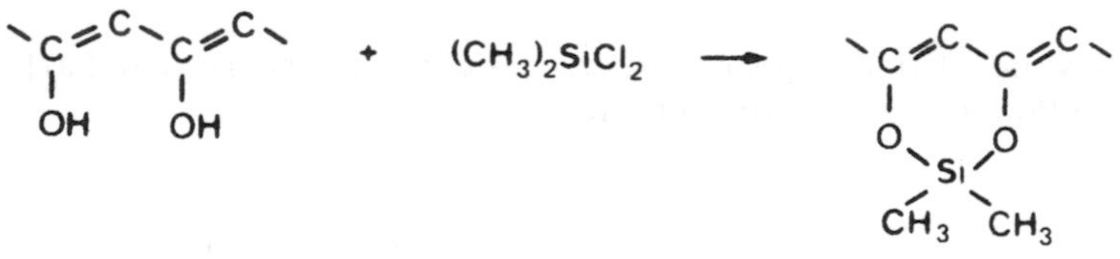

PK   PSiK

C. Systems Containing Nitrogen

Polythiazole

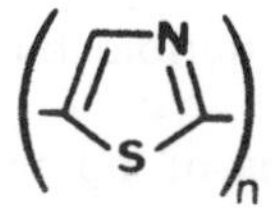

This material was prepared by chemical condensation. The reaction product is an infusible powder which decomposes at temperatures

above 240°C. A conductivity of 10^{-7} S/cm is attained with iron (III) chloride [105].

Polytriazoles

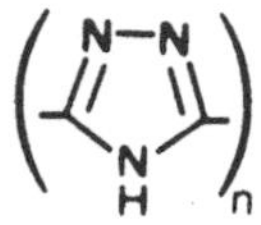

These products, preferably with substituted N_1, are extremely stable electroactive polymers. They undergo reversible oxidation and reduction [106].

Polyoxidiazoles

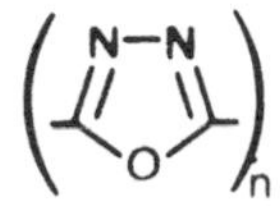

These are extremely heat-stable products and attain conductivities after oxidation or reduction of about 10^{-1} S/cm [107].

Poly(azabicyclo[3,3,0]diene)

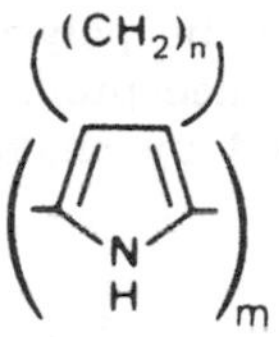

These polymers and their substitution products [109] are soluble and processable. After electrochemical oxidation, they attain conductivities of up to 100 S/cm.

Poly(tetrahydropyridyl)methylene

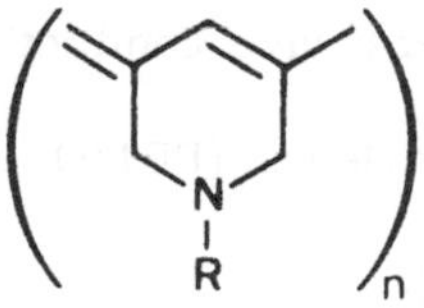

R=H, alkyl, $-COCH_3$

These polymers are prepared by polymerizing dipropargylamine and derivatives. Iodine doped conductivities of 10^{-3} S/cm are achieved [110a].

Polybenzopyrrolines [110]

These compounds are produced from o-dicyanobenzene. They are stable in air and can be converted by the usual oxidation or reduction agents into products with conductivities as high as 200 S/cm.

D. Systems Containing Sulfur

Polythianaphthene

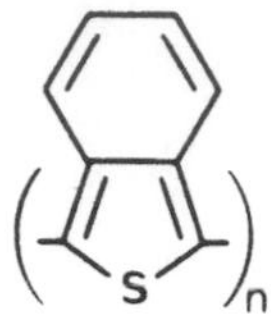

Polythianaphthene can be obtained by chemical or electrochemical poly-merization [113]. Conductivities of 10^{-2} S/cm are obtained with sul-furic acid. Similar polymers can be obtained if the starting material is dihydroisothianaphthene [114].

Dithienobenzene

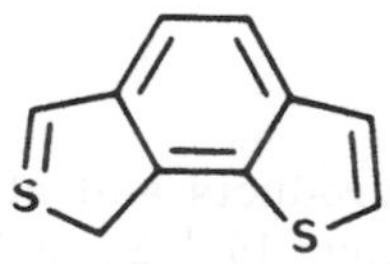

Dithienobenzene [115,121] can be polymerized electrochemically or chemically. It is an electron acceptor.

Similar behavior is displayed by analogous structures such as these or the corresponding dimers:

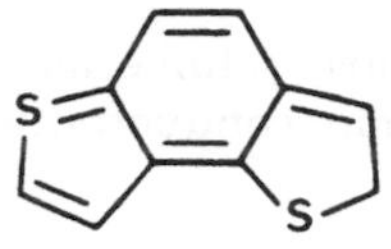

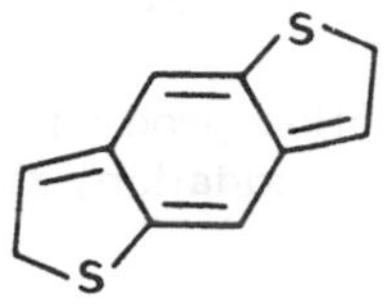

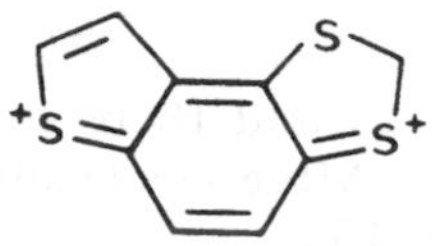

The oxidized polymers have conductivities of about 10^{-2} S/cm.

Polyphenylenethiovinylenes (PPTV)

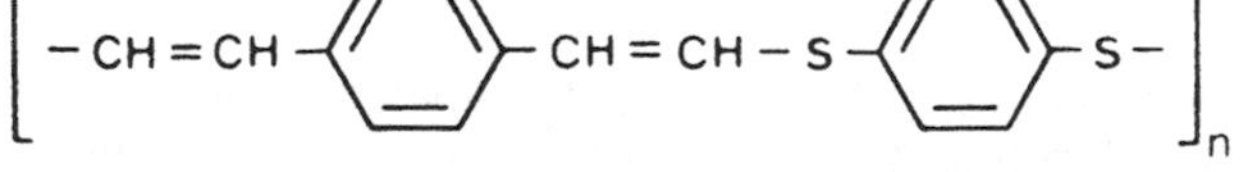

If 1,4-benzenedithiol and 1,4-diethynylbenzene are exposed to ultra-violet radiation at $-20°C$, they enter into addition reactions as shown [116]:

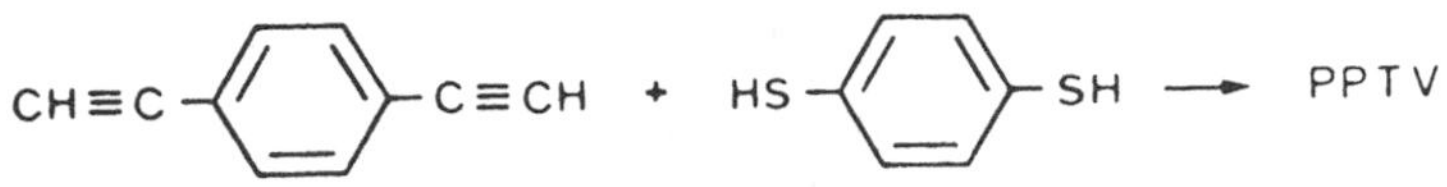

$$CH\equiv C - \langle\bigcirc\rangle - C\equiv CH \quad + \quad HS - \langle\bigcirc\rangle - SH \quad \longrightarrow \quad PPTV$$

The polymers are soluble and readily processable.

E. Heterobridged Annellated Thiophenes

2,5-Dithienylenesulphide and Cyclic Derivatives

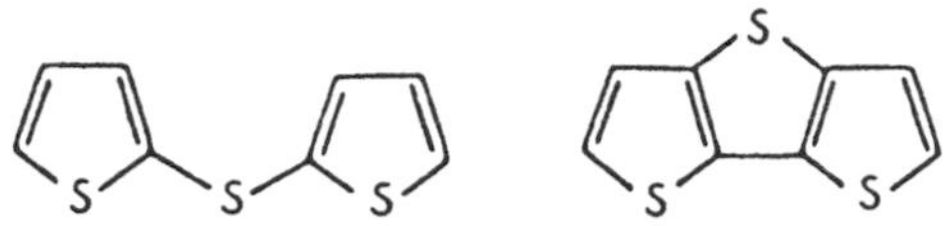

These products are readily accessible by condensation reactions and can be polymerized. High conductivities, i.e., 50 S/cm, can be attained with arsenic pentafluoride as the dopant [121].

Triphenoldithiazine (TPDT)

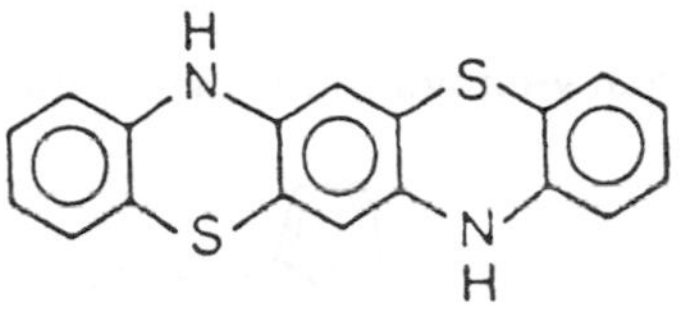

TPDT can be readily produced by reacting N,N'-diphenyl-p-phenyl-enediamine with sulfur [117]. This method appears to be a fundamental route for producing cyclic structures containing sulfur by thionation. In common with thianthrene or its analogues [118], TPDT can be reversibly oxidized and attains conductivities of about 10^{-3} S/cm.

Lignin —S— Compounds

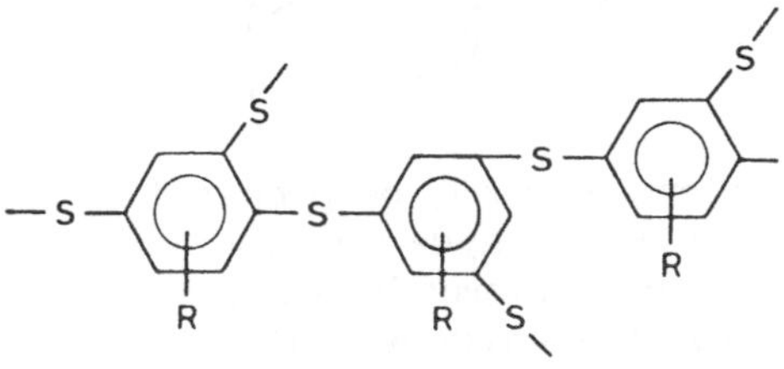

Modifying lignin with sulfur gives rise to heat-stable polymers that may attain conductivities of 10^{-3} S/cm, e.g., with iron(III) chloride [119].

Conductive Pendant Groups

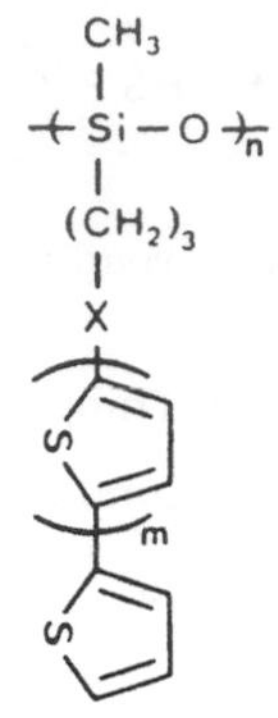

X = nil, —O—, —S—

Polysiloxanes have been synthesized with oligothiophene side groups.
The polymers yield deep-black CT-complexes with TCNQ [120].

F. Heteroarylenevinylene Structures

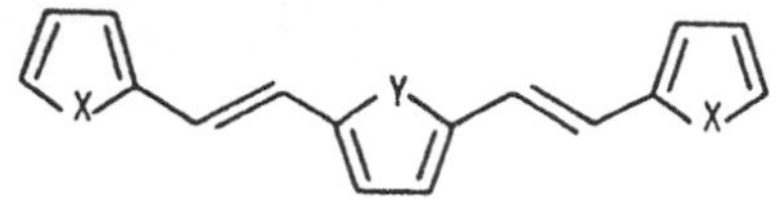

X=NMe; Y=S
X=Y=NMe
X=S; Y=NMe

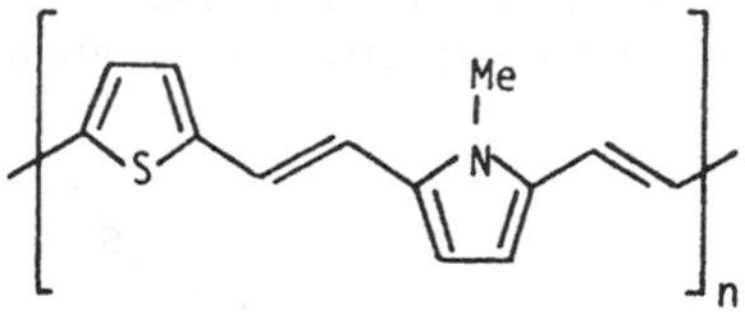

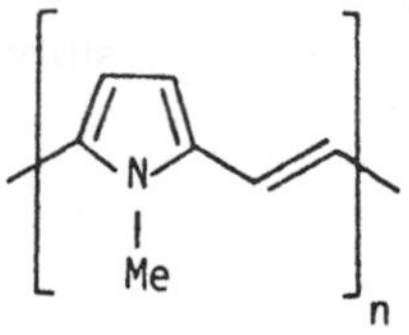

These compounds may be prepared by the condensation of hetero-
arylene-2,5-dialdehyde with the suitable diethylheteroaryl-2-methane
phosphonate. Alternatively, they can be produced from the proper
heteroaryl-2-aldehyde with the appropriate tetraethylheteroarylene
2,5-dimethane phosphonate. Polymeric heteroarylenevinylene materials
may also be produced by oxidative coupling or condensation via
lithiated intermediates. Conductivities of 3 S/cm can be attained by
iodine oxidation [121,122].

  Polymeric phthalocyanines have been extensively studied since the
mid-1950s. They have been comprehensively described by Hanack
[123].

The phthalocyaninephosphazenes are a new type. These compounds synthesized by Allcock [124,125], reach conductivities of up to 10^{-3} S/cm after they have been oxidized with iodine.

R=H   R=CH$_3$

R=CH$_2$OC$_6$H$_5$ R=CH$_2$OCH$_2$CH$_2$OCH$_3$

Wöhrle [126] has described modified polystyrenes with porphine or phthalocyanine side groups. These systems display catalytic and photovoltaic effects.

Itoh [132] has reported on analogous polymers with a methacryl backbone.

Polymers with porphine units are also obtained if pyrroles are polymerized electrochemically in the acid electrolyte together with other aldehydes, e.g., pyrrole aldehyde, thiophene aldehyde or other compounds with active CO groups [128,129].

G. Other Electrically Conducting Polymeric Systems

Bimetallic Complexes: Complexes of the type $PdPt(C_2S_2O_2)_2^-{}_n$ have conductivities on the order of 10^{-2} S/cm without doping. These complexes, which contain dithiooxylato bridges, are obtained as black powders in water by reacting the thioxalate $K_2Pd(C_2S_2O_2)_2$ with potassium tetrachloroplatinate K_2PtCl_4 [127].

Nickel Thioxolate Polymers: These polymers have conductivities of 20 S/cm [131].

Polydecker Sandwich Complexes: These interesting structures are obtained by polycondensation reactions. Their conductivities are less than 10^{-2} S/cm [130]. They are synthesized as follows:

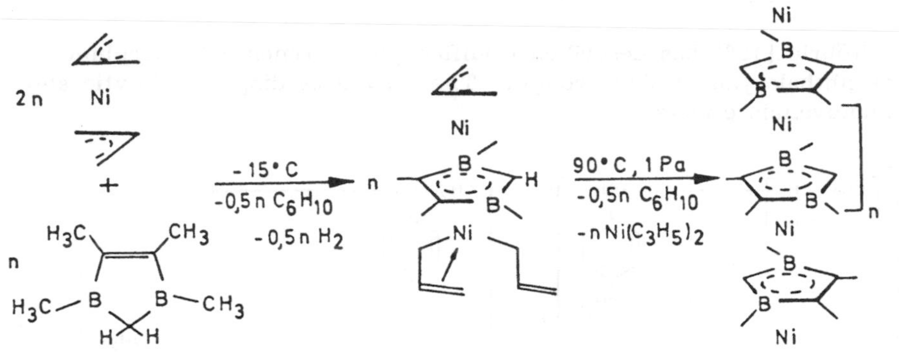

Polyamides Containing Tetrathiafulvalene Moieties in the Main Chain. The original structural principles affords considerable scope for variation [138]. These polymers are synthesized according to the following scheme.

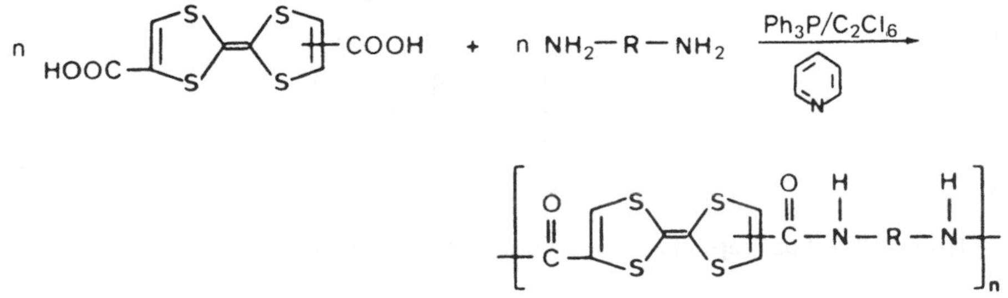

The conductivities after oxidation with bromine are of the order of 10^{-5} S/cm.

VI. HETERO π- AND σ-SYSTEMS

In this chapter, a few examples of hetero π-systems made conductive by oxidation or reduction are presented. This does not concern ionic conductivity [133], as is observed with the formation of complexes of polysiloxanes and lithium perchlorate, which have ionic conductivities of about 10^{-4} S/cm.

A. Polysilanes

Polysilanes are insulators but become semiconducting upon treatment with oxidizing agents such as antimony or arsenic pentafluoride. Typical conductivity data [137] are as follows:

$$[(Me_2Si)_{1.0}(PhMeSi)_{1.0}]_n \qquad\qquad 10^{-6} \text{ S/cm}$$

$$[(Me_2Si)_{0.5}(PhMeSi)_{1.0}]_n \qquad\qquad 10^{-1} \text{ S/cm}$$

(both doped with arsenic hexafluoride)

The mechansim of conductivity is somewhat different from that for other conducting polymers, since it depends on delocalization of electrons rather than on π electrons. The charge appears to be mobile along the length of the polysilane chain.

Unlike their saturated carbon backbone counterparts, the polysilanes show an intense electronic absorption described variously as either a $\sigma-\sigma$ or a $\sigma-3d\pi$ transition [134,135].

Polysilanes are just as unstable in air and moisture as the other conductive polymers mentioned in the literature.

B. Polyphosphazene

Polyphosphazene electroactive materials have recently emerged, either as electronic conductors or as solid electrolytes for light-weight rechargeable batteries [136]. Polyphosphazene is the most interesting of the solid electrolyte polymers.

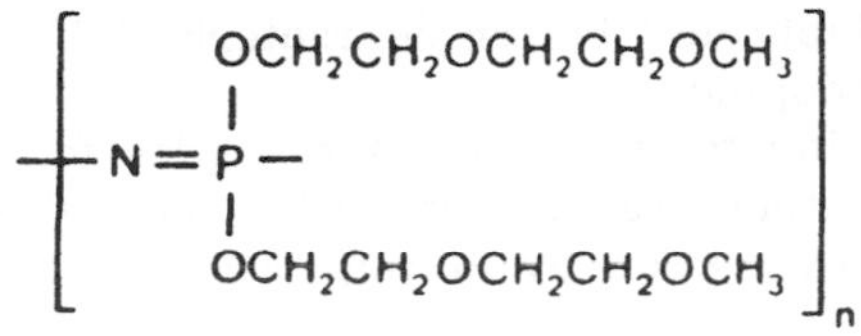

$$\left[\begin{array}{c} OCH_2CH_2OCH_2CH_2OCH_3 \\ | \\ -N=P- \\ | \\ OCH_2CH_2OCH_2CH_2OCH_3 \end{array} \right]_n$$

Solutions of lithium triflate in this polymer have ionic conductivity
at room temperature nearly three orders of magnitude higher than
that of similar systems formed from totally organic polymers.

VII. CONSIDERATIONS

In his notes, Samuel Taylor Coleridge used the word "nimiety" to
describe the results of digesting a literary glut, particularly of
secondary publications, duplicating from Adam and Eve. There has
been a veritable flood of recent literature on electrically conductive
polymers (ECPs), and it has been difficult to present a complete
picture of all that is significant. Therefore, the only contributions
considered in this summary are those that—in our opinion—concern
new and stimulating structures, syntheses, and concepts that have
arisen in the last three years.

 We realize that it is presumptious on our part to put forward suggestions for diverting the flood fo scientific publications on ECPs to
allow an easy review. However, in our opinion, the following points
are important and we hope all authors will consider them.

1. The *polymerization process* should be clearly defined, and
   the synthesis conditions should be clearly indicated, including
   the purity of the starting materials, the preparation of the
   catalysts, and the purification of the samples.
2. The product obtained should be *characterized*, e.g., by
   elementary analysis, particularly of the impurities, or by
   spectroscopy (IR, NMR, etc.)
3. *Conductivity measurements* of specimens prepared under defined conditions should be determined by the four-probe method
   under argon atmosphere.
4. *Doping* by p-oxidation or n-reduction should be carried out
   under standard conditions, e.g., with iodine saturated in carbon tetrachloride or with sodium/naphthalene in tetrahydrofuran.
5. *Stability*: The specimens should be kept in air in a standard
   laboratory atmosphere, and the oxygen content and the conductivity should be determined over time, e.g., 1 day, 1 week,
   1 month. The results should be expressed as a fraction of the
   original values: O_0/O_t or σ_0/σ_t (O = oxygen content, σ =
   conductivity S/cm).

ECPs could be evaluated more reliably and better characterized if
proposals along these lines were adopted.

 A study of the literature will bring forth sufficient examples for
differences in evaluation. Some authors regard 10^{-4} S/cm as highly
conductive and substantiate their evaluation by stating that this value

is, after all, 10,000 times higher than the original. Other authors regard 10^2 S/cm as highly conductive, and still others consider that an ECP is not very conductive unless the figure is 20,000 S/cm or higher.

The significance of indicating the synthesis conditions in detail can be readily appreciated from the work of Tanaka et al. [11]. Because of the extreme purity of the starting materials and the extreme conditions adopted for purification, a $(CH)_x$ with novel properties was obtained. The same applies to the $(CH)_x$ polymerized at room temperature in a viscous medium, e.g., silicone oil, or to polymerizing with heat-treated catalysts.

Another example is electrochemical polymerization. In this case, the polymer's properties depend on the electrode material, its surface, the electrolytes, the counterions, temperature, etc. Other important factors are pH and the purity of the monomers, as in the production of polyaniline.

Many workers engaged in the ECP field now appreciate the importance of knowing details of the synthesis parameters and the characteristics of the materials, including those that appear to be irrelevant. For instance,

$$(-CH=CH-)_m \quad \text{or} \quad \left(\underset{\underset{H}{N}}{\bigsqcup} \right)_n \quad \text{or} \quad \left(\bigcirc \right)_o \quad \text{etc}$$

are not automatically identical to

$$(-CH=CH-)_x \quad \text{or} \quad \left(\underset{\underset{H}{N}}{\bigsqcup} \right)_y \quad \text{or} \quad \left(\bigcirc \right)_z \quad \text{etc.}$$

It can be concluded that we are on the right track. Considerable and encouraging success in the synthesis of novel and improved materials has been achieved in a variety of ways, and many novel applications have emerged. This success is bound to lead to an even higher level of activity in years hence.

REFERENCES

1. W. J. Feast, *Handbook of Conducting Polymers*, Vol. 1, T. A. Skotheim, ed., Marcel Dekker, New York, 1986, pp. 1–43.
2. G. Wegner, *Angew. Chem.*, *93*:360 (1981).

3.   H. Naarmann, *Angew. Makromol. Chem. 109*:295 (1982).
4.   H. Naarmann, *Synth. Met. 17*:223 (1987).
5.   H. Naarmann, N. Theophilou, *Synth. Met.*, *22*:1 (1987).
6.   H. Haberkorn, H. Naarmann, K. Penzien, J. Schlag, and
     P. Simak, *Synth. Met.*, *5*:51 (1982).
7.   BASF Germany, EP 88301, Mar. 5, 1982/Feb. 25, 1983,
     H. Naarmann, K. Penzien, and P. Simak.
8.   H. Shirakawa, Y. C. Chen, and K. Akagi, *Synth. Met.*, *14*:
     173 (1986).
8a. K. Akagi, S. Katayama, H. Shirakawa, K. Araya, A. Mukoh,
     and T. Narahara, *Synth. Met.*, *17*:241 (1987).
9.   M. Aldissi, *J. Polym. Sci. Polym. Lett. Ed.*, *23*:167 (1985).
10. U.S. Pat. 4468291, Jun. 27, 1983/May 28, 1984, BASF Germany
     H. Naarmann, G. Köhler, and J. Schlag.
10a. A. F. Diaz, K. K. Kanazawa, and G. P. Gardini: *J.C.S.
     Chem. Commun.*, 1979:635.
10b. H. Lund, *Elektroredaktioner i Organisk Polarografi Og
     Voltammetri*, Aarhus Stiftsbogtrykkerie, Aarhus, 1961.
11. H. Tanaka and T. Danno, *Synth. Met.*, *17*:545 (1987).
12. Agency of Ind. Sci. Tech., Jap. 196062 & Jap. 196063,
     Oct. 21, 1983/May 17, 1985.
13. Polyplastics KK, Jap. 253228, Nov. 30, 1984/June 11, 1986.
13a. Costello C. A. and McCarthy, T. J., A.C.S., Div. Polym. Chem.
     Polym. Preprints, 26:68 (1985).
14. Hitachi KK, Jap. 129408, Jun. 1, 1984/Jan. 23, 1985.
15. BASF Germany, DE-OS 3315854, Apr. 30, 1983/Oct. 31, 1984,
     H. Naarmann and G. Köhler.
16. H. Naarmann, N. Theophilou, submitted to *Synth. Met.*
17. N. Theophilou, A. Munardi, R. Aznar, J. Sledz, F. Schué,
     and H. Naarmann: *Eur. Polym. J.*, *23*:15 (1987).
18. N. Theophilou, A. Munardi, R. Aznar, J. Sledz, F. Schué,
     and H. Naarmann: *Eur. Polym. J.*, *23*:11 (1987).
19. I. Kminek and I. Trekoval, *Makromol. Chem. Rapid Commun.*,
     7:53 (1986).
20. N. Theophilou: Ph.D. thesis, Univ. Languedoc, Montpellier,
     France, 1985.
20a. N. Theophilou, R. Aznar, A. Munardi, J. Sledz, and F. Schué,
     J. Macromol. Sci. Chem., *A24*:797 (1987).
21. H. Haberkorn, W. Heckmann, G. Köhler, H. Naarmann, P. Simak,
     N. Theophilou, and R. Voelkel; submitted to Eur. Polym. J.
22. D. Meziane, A. Soum, and M. Fontanille, *Makromol. Chem.*,
     186:367 (1985).
23. R. N. Leyden, U.S. Pat. 4588609, Nov. 26, 1984/May 13, 1986.
24. F. Schiller Univ. Jena, DD Pat. 270439, Dec. 7, 1984/Nov. 6, 1985.
25. F. E. Karasz, J. D. Capistran, and D. R. Gagnon, *Mol. Cryst.
     Liq. Cryst.*, *118*:327 (1985).
26. I. Murase, *Mol. Cryst. Liq. Cryst.*, *118*:333 (1985).

27. F. S. Bates and G. L. Baker, *Macromolecules, 16*:707 (1983).
28. S. Destri, M. Catellani, and A. Bolognesi, *Makromol. Chem. Rapid Commun., 5*:353 (1984).
29. M. Aldissi and A. R. Bishop, *Polymer, 26*:662 (1985).
29a. F. Schue, R. Aznar, A. Munardi, J. Sledz, L. Giral, and P. Bernier, *Synthesis of Block Copolymers*, ICSM 84, Abano Terme, Italy, 1984.
30. M. E. Galvin and G. E. Wnek, *Polymer Bull., 13*:109 (1985).
31. BASF Germany, DE 3312300, April 6, 1983/Oct. 11, 1984, H. Naarmann, G. Köhler, and J. Schlag.
32. ELF Aquitaine, DE-OS 3443202, Oct. 27, 1984/Jun. 6, 1985, R. Gouarderes and G. Merceur.
33. BASF Germany, EP 129070, May 25, 1983/Dec. 27, 1984, G. Wegner and W. Wernet.
33a. H. Naarmann, *Electrochemistry of Polymer Layers*, Internat. Workshop, Duisburg, FRG, Sept. 15, 1986.
33b. BASF Germany, DE-OS 3518886, May 25, 1985/Nov. 27, 1986, H. Naarmann.
33c. BASF Germany, DE-OS 3421296, June 8, 1984/Dec. 12, 1985, H. Naarmann and H. Münstedt.
33d. BASF Plastics, Research and Development, BASF, FRG, Oct. 1986.
33e. BASF Germany, DE-OS 3508266, Mar. 8, 1985/Sept. 11, 1986, H. Naarmann.
34. M. Velazquez-Rosenthal, T. A. Skotheim, and C. A. Linkous, *Synth. Met., 15*:219 (1986).
35. BASF, DE-OS 3425511, Jul. 11, 1984/Jan. 16, 1986, H. Naarmann and G. Köhler.
36. A. Heeger and F. Wudl: *Electrochemistry of Polymer Layers*, Internat. Workshop, Duisburg, FRG, Sept. 15, 1986.
36a. A. O. Patil, Y. Ikenoue, F. Wudl, and A. Heeger, *J. Am. Chem. Soc., 109*:1858 (1987).
37. BASF Germany, DE-OS 3246319, Dec. 15, 1982/Jun. 20, 1984, H. Naarmann and G. Köhler.
38. W. R. Grace & Co., USA EP 145843, Dec. 14, 1983/Jun. 26, 1985, R. W. Peyton.
39. Toray Ind. Inc., Jap. 004726, Jan. 17, 1984/May 7, 1985.
40. BASF Germany, EP-OS 98988, Jun. 16, 1983/Jan. 25, 1984, H. Naarmann, G. Köhler, and J. Schlag.
41. A. J. Diaz and T. J. McCarthy, *Macromolecules, 18*:869 (1985).
42. G. Lugli, U. Pedretti, G. Perego: *J. Polym. Sci. Polym. Lett. Ed., 23*:129 (1985).
43. L. A. Schulz and W. P. Roberts, A.C.S., *Div. Polym. Chem. Polym. Preprints, 25*:253 (1984).
44. K. Muaro and K. Suzuki, A.C.S., *Div. Polym. Chem. Polym. Preprints, 25*:260 (1984).
45. M. Aldissi, A.C.S., *Div. Polym. Chem. Polym. Preprints, 26*: 269 (1985).

46.   T. Yamamotu, K. Osakada, T. Wakabayashi, and A. Yamamoto, *Die Makromol. Chemie Rapid Commun.*, *6*:671 (1985).

47.   K. Soga, Y. Kobayashi, S. Ikeda, and S. I. Kawakami, *J.C.S. Chem. Commun.*, 1980:931.

48.   M. Satch, M. Tabata, K. Kaneto, and K. Yoshino, *Polym. Commun.*, *26*:356 (1985).

49.   T. Ojio and S. Miyata, *Polym. J.*, *18*:95 (1986).

50.   BASF Germany, EP 0053669, Oct. 8, 1981/Feb. 6, 1985, H. Naarmann, K. Penzien, D. Naegele, and J. Schlag.

51.   Schinetsu Chem. Ind., Jap. 014088, Jan. 28, 1984/Aug. 19, 1985.

52.   Agency of Ind. Sci. Tech., Jap. 251519, Aug. 15, 1985/Jun. 5, 1986.

53.   Olympus Optical, Jap. 222938, Nov. 26, 1983/Jun. 21, 1985.

54.   Agency of Ind. Sci. Tech., Jap. 022389, Feb. 9, 1984/Aug. 29, 1985.

55.   Mitsubishi Chem. Ind., Jap. 041771, Mar. 5, 1984/Sept. 24, 1985.

56.   T. Tomura, Jap. 085278, Apr. 27, 1984/Nov. 15, 1985.

57.   Shinetsu Chem. Ind., Jap. 243609, Dec. 22, 1983/July 18, 1985.

58.   J. E. Frommer, R. L. Elsenbaumer, H. Eckhardt, and R. R. Chance, *J. Polym. Sci. Polym. Lett. Ed.*, *21*:39 (1983).

59.   Mitsubishi Chem. Ind., Jap. 069856, Apr. 7, 1983/May 15, 1985.

60.   Asahi Chem. Ind., Jap. 194351, Oct. 19, 1983/May 15, 1985.

61.   H. W. Gibson, *Handbook of Conducting Polymers*, Vol. 1, T. A. Skotheim, ed., Marcel Dekker, New York, 1986, p. 405.

62.   Bridgestone Tire KK, Jap. 225600, Nov. 30, 1983/Jun. 26, 1985.

63.   G. Bidan, P. Audebert, M. Lapkowski, and D. Limosin, *Synth. Met.*, *19*:1010 (1987).

64.   J. R. Reynolds, P. A. Poropatic, and L. Toyooka, *Synth. Met.* *18*:95 (1987).

65.   CNRS CENT. NAT Recherche, EP 163559, Apr. 20, 1984/Dec. 4, 1985, I.C. Moutet, G. Bidan, and A. Deronzier.

66.   Goodrich BF Co., U.S. Patent 4543402, Apr. 18, 1983/Sept. 24, 1985, L. Traynor.

67.   A. I. Nazzal and G. B. Street, *J. Chem. Soc. Chem. Commun.*, *6*:375 (1985).

68.   Showa Denko KK, Jap. 123852, Jun. 18, 1984/Jan. 10, 1986.

69.   Toyota Cent. Res. & Dev., Jap. 162174, Jul. 31, 1984/Feb. 26, 1986.

70.   Polaroid Corp., EP 160207, Mar. 23, 1985/Nov. 6, 1985.

71.   S. Tanaka, K. Okuhara, and K. Kaeriyama, *Makromol. Chem.*, *187*:2793 (1986).

72.   Masa-aki Sato, S. Tanaka, and K. Kaeriyama, *Synth. Met.*, *18*:229 (1987).

73. F. Chao, *Ann. Phys.* (Paris), *104*:11 (1986); CA., Vol. 104, 158130t (1986); F. Garnier, *J. Mater. Sci.*, *20*:2687 (1985).

74. Shinko Kagaku Kogyo, Jap. 158536, Aug. 30, 1983/Mar. 19, 1985.

75. Matsushita Elec. Ind., Jap. 008039, Jan. 19, 1984/Aug. 10, 1985.

76. Thomson-CSF, Fr. Pat. 2554133, Oct. 28, 1983/May 3, 1985.

77. Chevron Research Co., U.S. Pat. 4505844, Nov. 17, 1982/ Mar. 19, 1985, P. Denisevich.

78. Chevron Research Co., U.S. Patent 4620943, Oct. 25, 1984/ Nov. 4, 1986, S. Suzuki, P. Denisevich, A. H. Schroeder, and V. P. Kurkow.

79. Honneywell Inc., U.S. Pat. 4548738, Jun. 29, 1984/Oct. 22, 1985, U.S. Pat. 4568482, Aug. 23, 1983/Feb. 4, 1986, S. A. Jenekhe and B. J. Fure.

80. Showa Denko, Jap. 206141, Oct. 3, 1984/May 1, 1986.

81. N. Oyama and T. Ohsaka, *Synth. Met.*, *18*:375 (1987).

81a. M. Angelopoulos, S. P. Ermer, A. Ray, and A. G. McDiarmid; 193rd ACS National Meeting, Denver, Colorado, April 9, 1987, I&EC Section, No. 46.

82. BASF Germany, DE-OS 3338904, Oct. 27, 1983/May 9, 1985, H. Naarmann, P. Simak, and J. Schlag.

83. C. Taliani, R. Danieli, R. Zamboni, P. Ostoja, and W. Porzio, *Synth. Met.*, *18*:177 (1987).

84. D. Mukherjee, P. Pramanik, and B. K. Mathur: *Polym. Commun.* *27*:218 (1986).

85. R. Lazzaroni, A. DePrijck, J. Riga, J. J. Verbist, J. L. Brédas, J. Delhalle, and J. M. Andre, *Synth. Met.*, *18*:123 (1987).

86. K. Tamura, T. Masuda, and T. Higashimura, *Polym. J.*, *17*: 815 (1986).

87. Tech. Hochschule Karl Marx, GDR, DD 216939, Aug. 11/1983/ Jan. 2, 1985.

88. Y. Huang, J. Li, Q. Zhou, and J. Zhou, *J. Polym. Sci. Polym. Chem. Edit.*, *23*:1853 (1985).

89. Y. Tanaka, H. Matsuda, S. Iijima, H. Fujikawa, H. Nakanishi, M. Kato, and S. Kato: *Die Makromol. Chem. Rapid Commun.*, *6*:715 (1985).

90. K. Akagi, M. Nishiguchi, H. Shirakawa, Y. Furukawa, and I. Harada, *Synth. Met.*, *17*:557 (1987).

91. BASF Germany, DE-OS 3344753, Dec. 10, 1983/Jun. 20, 1985, H. Naarmann, and G. Köhler.

92. W. I. Feast and K. Harper, *Brit. Polym. J.*, *18*:161 (1986).

93. BASF Germany, EP-OS 111237, Nov. 28, 1983/Jun. 20, 1984, H. Naarmann and G. Köhler; U.S. Pat. 4435039, Apr. 4, 1982/ Jun. 7, 1984.

94. I. Schopov and L. Mladenova, *Makromol. Chem., Rapid Commun.*, *6*:659 (1985).

95.   H. Bräunling and R. Jira, *Syn. Met.*, *17*:691 (1987).

96.   BASF Germany, DE-OS 3410494, Mar. 22, 1984/Oct. 3, 1985,
      G. Köhler, H. Naarmann, and H. Münstedt.

97.   H. Mains and J. H. Day, *J. Polym. Sci. B*, *1*:347 (1963).

98.   S. Naitoh, *Synth. Met.*, *18*:237 (1987).

99.   U. Haya, P. N. Bartlett, and G. H. Dodd, *Polym. Commun.*,
      27:362 (1986).

100. D. Zhu, S. Shi, and R. Quian, *Macromol. Chem. Rapid Commun.*
      7:313 (1986).

101. S. T. Wellinghof, Z. Deng, J. Reed, J. Racchini, and S. A.
      Jenekhe, *ACS, Div. Polym. Chem. Polym. Preprints*, *25*:238
      (1984).

102. Shin-Gijutsu, Jap. 60195124, Mar. 16, 1984/Oct. 3, 1985.

103. S. Mizogami and S. Yoshimura, *Synth. Met.* *18*:479 (1987).

104. H. A. Staab and I. Weiser, *Angew. Chem.*, *96*:602 (1984).

105. A. Bolognesi, M. Catellani, S. Destri, and W. Porzio, *Synth.
      Met.*, *18*:129 (1987).

106. Chevron Research Co., U.S. Pat. 4519940, Nov. 17, 1982/May
      28, 1985; 4505842, Nov. 17, 1982/Mar. 19, 1985.

107. Furukawa Electric Co., Jap. 151163, Aug. 19, 1983/Mar. 9,
      1985.

108. Y. Okamoto, E. F. Hwang, and M. C. Wang, *J. Polym. Sci.
      Polym. Lett. Edit.*, *23*:285 (1985).

109. B. F. Goodrich Co., U.S. Pat. 4487667, May 16, 1983/Dec.
      11, 1984, L. Traynor.

110. Showa Denko, Jap. 207043, Oct. 4, 1984/May 1, 1986.

110a. W. R. Grace & Co., U.S. Pat. 4511702, Apr. 10, 1984/April
      16, 1985.

111. Showa Denko, Jap. 206142, Oct. 3, 1984/May 1, 1986.

112. BASF Germany, DE-OS 3409655, Mar. 16, 1984/Dec. 12, 1985,
      J. Nickl, H. Naarmann, and H. Möhwald.

113. Univ. of California, U.S. Pat. 736984, May 22, 1985/Dec. 18,
      1985, F. Wudl, M. Kobayashi, and A. Heeger.

114. BASF Germany, DE-OS 3502937, Jan. 30, 1985/Jul. 31, 1986,
      H. Naarmann and G. Köhler.

115. BASF Germany, EP-OS 176921, Sept. 24, 1985/Apr. 9, 1986,
      H. Naarmann and G. Köhler.

116. E. Kobayashi, T. Ohashi, and I. Furukawa, *Macromol. Chem.*,
      187:2525 (1986).

117. C. Della Casa, P. C. Bizzarri, and F. Adreani, *Synth. Met.*,
      18:381 (1987).

118. BASF Germany, DE-OS 352 102, Jun. 5, 1985/Dec. 11, 1986,
      S. Hünig and H. Naarmann.

119. K. Levon, A. Kinanen, and I. I. Lindberg, *Polym. Bull.*, *16*:
      4:433 (1986).

120. G. E. Wnek and S. A. Arnold, *ACS, Proc. Div. Polym. Mat.
      Eng.*, *53*:89 (1985).

121. A. Berlin, S. Bradamante, G. Pagani, and F. Sannicolo, *Synth. Met.*, *18*:157 (1987).

122. G. Pagani, A. Berlin, S. Bradamante, R. Ferraccioli, and F. Sannicolo, *Synth. Met.*, *18*:117 (1987).

123. M. Hanack, *Handbook of Conducting Polymers*, Vol. 1, T. A. Skotheim, ed., Marcel Dekker, New York, 1986, p. 133.

124. H. R. Allcock and T. X. Neenan, *Makromol.*, *19*:1495 (1986).

125. H. R. Allcock, *Chem. Eng. News*, March 18, 1985, p. 22.

126. D. Wöhrle and G. Krawczyk, *Makromol. Chem.*, *187*:2535 (1986).

127. C. Alvarez, I. Rodriguez, M. E. Lopez-Morales, R. Escudero, and J. Gomez-Lara, *Synth. Met.*, *19*:984 (1987).

128. BASF Germany, EP-OS 166980, Jun. 3, 1985/Jan. 8, 1986, H. Naarmann, G. Köhler, H. Hiller, and P. Simak.

129. BASF Germany, D. E. 3420854, Jun. 5, 1984/Dec. 5, 1985, H. Naarmann, G. Köhler, H. Hiller, and P. Simak.

130. U. Zenneck, Th. Kuhlmann, S. Roth, J. Roziere, W. Siebert, and J. Zwecker, *Synth. Met.*, *19*:757 (1987).

131. J. R. Reynolds, J. C. Karasz, and P. Lillya, ACS, *Div. Polym. Chem. Polym. Preprints*, *25*:242 (1984).

132. H. Itoh, S. Kondo, E. Masuda, K. Hanabusa, H. Shirai, and N. Nojo, *Makromol. Chem. Rapid Commun.*, *7*:585 (1986).

133. D. Fish, J. M. Khan, and J. Smid, *Makromol. Chem. Rapid Commun.*, *7*:115 (1986).

134. J. M. Zeigler and L. A. Harrah, ACS, *Proc. Div. Polym. Mat. Eng.*, *53*:486 (1985).

135. C. G. Pitt: "Homoatomic Rings, Chains and Macromolecules of Main-Group Elements," Rheingold A. L., ed., Elsevier Science Publishing, New York, 1977, p. 203.

136. H. R. Allcock, *Makromol. Chem. Macromol. Symp.*, *6*:118 (1986); P. M. Blonsky, *Polym. Mat. Sci. Eng.*, *53*:118 (1985).

137. R. West, *J. Organomet. Chem.*, *300*:327 (1986).

138. M. Watanabe, T. Iida, K. Sanui, N. Ogata, T. Kobayashi, and Z. Ohtaki, *J. Polym. Sci. Polym. Chem. Edit.*, *22*:1299 (1984).

2

Aprotic Polymer Electrolytes and Their Applications

Michel Gauthier / Institut de Recherche d'Hydro-Québec (IREQ), Varennes, Quebec, Canada

Michel Armand / LIES ENSEEG Domaine Universitaire, St-Martin d'Hères, France

Daniel Muller / Groupe Elf-Aquitaine, Lacq, France

I. INTRODUCTION

Polymers, or plastic materials, have pervaded most technical applications in the industrial sector. The characteristics that have assured them such success are mainly:

An outstanding combination of physical properties: mechanical resistance, shock resistance, flexibility, elasticity, deformability, fatigue resistance, and light weight

Their chemical properties, such as the absence of corrosion, resistance to aggressive reagents like powerful solvents, bases and acids, which are often incompatible with common materials (e.g., metals)

Their ready applicability in rapid low-cost industrial processes, including extrusion, injection, and calendering.

Plastics are also associated with electrical insulation and are widely employed as insulating materials in wires and cables, as insulators in capacitors and switches, and as protection in domestic appliances. It is only recently, in fact, and still only on a limited basis that they have been associated with conduction applications, where their unique qualities can be put to good advantage. This chapter summarizes the various types of conduction that are possible in plastics, then focuses on one particular category of conducting polymers, namely, aprotic polymer electrolytes, which generally consist of a metallic salt dissociated by a solvating aprotic polymer.

The rapid expansion of research and development on these new electrolytes in the last ten years has come about largely as a result of general concerns about energy generation and conservation throughout the world. In particular, it is the development of rechargeable lithium batteries using polymer electrolytes that provided the original impetus for the study of ion-conduction mechanisms in these materials. This was followed by other energy-related applications, mainly photo-electrochemical and electrochromic devices. The common denominator at this juncture was the need for an easy-to-prepare adhesive electrolyte, which can be used to build a thin-film all-solid device offering a large surface area. Apart from these common attributes, each polymer application has its own very specific requirements as far as the nature of ion-conducting species or the chemical and electrochemical compatibility with other components of the complete device are concerned.

These sometimes stringent requirements, especially for rechargeable lithium batteries, are taken into account throughout the following review of state-of-the-art research into new conducting electrolytes. A tentative assessment is made of the performances of the principal devices incorporating aprotic polymer electrolytes that have been developed so far.

II. ORIGIN OF ELECTRIC CONDUCTION IN POLYMERS

The term *plastic* or *polymer conductor* is used in practice to cover a range of materials that often differ considerably in nature and conduction properties. Figure 1 shows the various types of conductivity that can be induced by physical or chemical means in polymers consisting of ethylene-based monomers. The figure also presents a simplified illustration of the origin and nature of electronic or ionic conductivity.

A. Electronic Conductivity

Charged Polymers

The first conducting polymers were obtained by charging insulating polymers with solid particles up to a volume fraction sufficient to induce electrical conductivity through percolation of the dispersed conductive phase (Fig. 1a). The conductive materials selected for this purpose can be carbon blacks or various metallic particles, such as stainless steel fibers. These composite plastics are used mostly as electromagnetic or radiofrequency barriers, among other applications, or for their antistatic properties.

Intrinsically Conducting Polymers

More recently, a new category of electronic conducting polymers has been developed, the so-called "conjugated unsaturated polymers" [1,2], which are generally intrinsic semiconductors that can be given a

Fig. 1   Schematic of various types of conductivity in polymers.

metallic character by n-type (Na°, Li°) or p-type (AsF$_5$, Br$_2$, FeCl3) chemical doping (Fig. 1b). These organic metals offer many possible applications, although chemical stability and processing problems have somewhat delayed their development. The use of polyacetylene and polypyrroles is now being investigated for their potential as electrode materials or as collectors in various types of battery [3,4].

B. Ionic Conductivity

Charged Polymers

Charged polymers are salt-bearing solvating polymer complexes which, since the work of Wright [5,6] on this topic, are known to be solid solutions of a salt in the polymer. The starting point for Wright's research was polyethylene oxide (PEO), which forms solid-state complexes with various alkaline salts as a result of cation solvation by the polymer. The same author also pointed out that these complexes can have precise stoichiometries and that they are ionically conducting (see Fig. 1c). It is these complexes which triggered the recent development of aprotic polymer electrolytes, now being studied in various electrochemical devices, including rechargeable lithium batteries.

Other charged complexes of the type polymer + salt + solvent have been known longer. In these complexes, however, the polymer does not necessarily contribute to the conductivity but tends to take the role of a gelling agent, with the conductivity provided by the salt dissolved in the liquid electrolyte. These systems are sometimes applied in aqueous-electrolyte batteries and have also been proposed for lithium batteries [7]. Their electrochemical performance is essentially the same as that of liquid electrolytes.

Polyelectrolytes

Polyelectrolytes are polymers that bear their own ion-generating groups chemically bound to the macromolecular chain; the presence of a counter-ion maintains the electroneutrality of the salt. A well-known example of polyelectrolytes is polystyrene sulfonate, which is used as an ion-exchange resin. More recently, perfluorinated Nafion type cationic polyelectrolytes have been developed and are now finding applications as membranes for brine electrolysis and devices such as fuel cells.

In the dry state, these polymers are poor conductors ($<10^{-12}$ S/cm) because the polar groups form stable, barely dissociated agglomerates. However, once these are swollen with an appropriate solvent, they can show quite significant conductivity (Fig. 1d). Nafion electrolytes swollen with propylene carbonate (PC) have been proposed as thin-film electrolytes [8] but their seems little hope of using them for lithium batteries because the perfluorinated backbone can be totally

reduced by the alkali metal. The reduction products are probably
mixed conductors, which makes it possible for the in-depth reaction
to propagate until complete reduction is attained [9].

Polyelectrolytes can, in principle, be modified to offer greater and
more specific electrical conductivity. Increasing the solvating capacity of the medium and selecting dissociable ion-bearing groups
could produce self-ionizing polymers, and investigations into purely
cationic conductors have already begun. This represents another
step forward in the effort to improve the nature and value of the
conductivity of polymer electrolytes currently made by complexing
a metal salt with a solvating polymer.

III. APROTIC POLYMER ELECTROLYTES AND
RECHARGEABLE LITHIUM BATTERIES

A. Recent Developments in Aprotic Polymer Electrolytes

Electrolytes consisting of solid solutions, that is, solvating aprotic
polymers containing metallic salts, fall into the category of charged
polymers (see Section II, part B). Among other research workers,
Fenton and Wright as early as 1973 reported the existence of crystalline complexes formed by polyethylene oxide (PEO) and various alkaline salts and pointed out that such complexes had a significant degree of conductivity. Since then, Armand has suggested that
aprotic electrolytes could be used in all-solid lithium batteries [10,11],
launching an idea that has stimulated an ever-growing number of
studies and development projects focused especially on polyethers derived from ethylene oxide and on their possible application in
batteries.

There are several reasons for such popularity. From the chemical
standpoint, inertia of the ether function, C—O—C, is an important
property for liquid organic electrolytes; it is known, for example,
that some cyclic ethers allow the electrochemical redeposition of
lithium [12]. Polyethers such as linear PEO show a higher stability
of the ether functions than cyclic ethers due to the absence of
strains [13] and, what is more, this stability is favored kinetically
by the solid state and by the absence of convection.

The PEO chain (CH_2—CH_2—O) seems to represent the best com-
promise between a high density of donor heteroatoms and a sufficiently long hydrocarbon chain segment. This gives the EO chain enough
flexibility to allow cation solvation by a cage effect, which involves
several donor heteroatoms. Neither polyoxymethylene, $(CH_2—O)_n$,
which is rich in donor groups but has a more rigid chain, nor poly-
oxacyclobutane, $((CH_2)_3—O)_n$, with its flexible chain but more widely
dispersed donor groups, possesses solvation properties equivalent to
those of the PEO chain. Moreover, they do not appear to form
complexes with alkali metals [14].

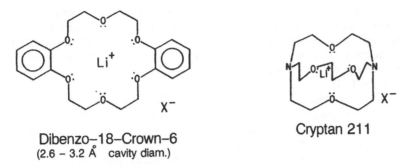

Dibenzo–18–Crown–6
(2.6 – 3.2 Å cavity diam.)

Cryptan 211

Fig. 2   Example of a crown ether and a cryptand.

The solvation capacity of PEO chains is due to the multidentate nature of intramolecular solvation, enhanced by entropy, and is also a result of the chain's ready ability to adopt a cage configuration in which the donor electron doublets of oxygen are directed inward [14,15]. This solvation mechanism is well known for crown ethers, or cryptands, in which cages of synthetically predetermined size can specifically solvate cations of the corresponding radius as shown in Fig. 2.

The advantage of the linear chains of PEO in the fabrication of a polymer electrolyte is that they can form solvation cages that can self-adjust fairly well to the size of the cation. This solvation mechanism, which is at least partially intramolecular, is also reversible and can take into account the presence of ion pairs or even triplets, quadruplets or other complex ions. Various degrees of participation of EO chains in cation solvation are therefore possible, as can be seen in Fig. 3.

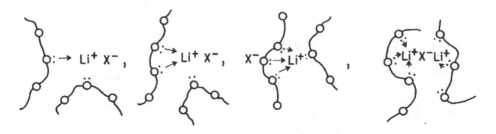

Fig. 3   Illustration of conformational flexibility of linear polyether enhancing ion solvation.

The diversity and reversibility of the various possible conformations of PEO chain segments around the cations also allow the latter to migrate from one solvating site to another, as discussed in section IV. By contrast, cationic conductivity is not found in exclusively anion-conducting crown ethers, where the cation is trapped in a solvating cage [16].

Developments in recent research into polyethers have led to other cation-conducting polymers being proposed, such as nitrogen [8,17, 18] or sulfur-based [19] solvating polymers, but these materials have a complex chemistry and their conductivity performance has been disappointing so far (see Fig. 9). Other solvating polymer/alkaline salt complexes are known, which comprise polar groups that are different from ethers [20–22]. These materials have still not been tested in Li batteries and appear to be destined rather for other electrochemical applications.

B. Solid-State Thin-Film Batteries with Ion-Insertion Electrodes

Lithium-ion conducting polymers are now available which can be used to fabricate thin-film electrolytes possessing a unique combination of characteristics: (a) solid state, (b) good adhesion, (c) deformability and (d) elasticity. As shown by Armand [10,11], these electrolytes can be usefully incorporated in solid-state batteries employing Li-reversible electrodes in electrochemical cells of the type:

$$(-) \text{ Li source / electrolyte / Li sink } (+) \tag{1}$$

Figure 4 presents a schematic illustration of a battery corresponding to the underlying principle of this electrochemical cell.

In order to achieve a high energy-to-weight ratio, strips of metallic lithium (1 faraday for 7 g) or of lithium-rich alloys are used as the negative electrode, while the positive electrode usually consists of a lithium insertion material in a powder form with a polymer electrolyte as the binding agent. The selection of host structures formed from lightweight elements that insert lithium ions at high voltages, such as TiS_2 and V_6O_{13}, give these batteries high theoretical specific energies of >500 Wh/kg. Figure 11, section V, presents the general characteristics of a polymer electrolyte thin-film battery with a TiS_2-positive electrode.

The polymer electrolyte has a triple role to play in this type of battery: (a) Li^+ ion carrier, (b) separator, and (c) binding agent for the positive electrode. In the first of these roles, the lithium salt present in the electrolyte does not participate globally in the electrode reactions, as is the case with conventional lead-acid or nickel-cadmium rechargeable batteries. Consequently the amount of electrolyte and the interelectrode spacing can be reduced to a strict

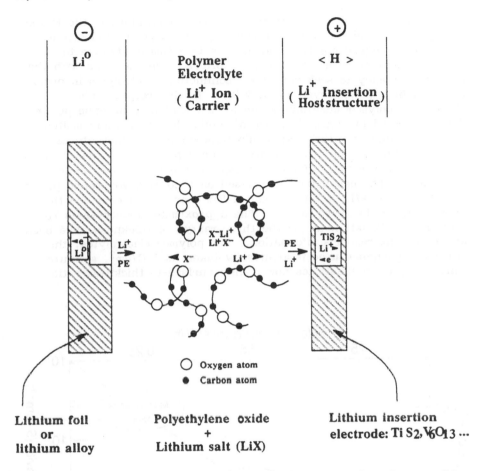

Fig. 4   Schematic diagram of a polymer electrolyte with Li reversible electrodes.

minimum.   Technically achievable with plastic materials, these reductions could greatly improve the ratio between the practical and the theoretical energy content of the battery in Fig. 4.

In the second role, the polymer electrolyte acts as an interelectrode separator, and such characteristics of the solvating polymer as high molecular mass and crosslinking provide a means of controlling the mechanical properties.

Thirdly, the polymer electrolyte is also present in the positive electrode (see Fig. 11), where it plays a crucial role as a binding agent for the composite electrode. As such, it allows all contacts and interfaces to be preserved, despite the many changes in volume

that occur during cycling. It also ensures a continuum in the transport properties of the separator right up to the positive electrode.

   The technological ability to fabricate a thin-film electrolyte to
serve exclusively as a Li^+ ion carrier offers a way to break with the
present tendency to seek very good conducting electrolytes in order
to attain high power levels (>100 W/L). In this respect, Fig. 5
shows how polymer electrolytes can in fact lead to volumetric power
densities equal in volume to those of liquid electrolytes and molten
salts. It illustrates the relationship between the conductivity of
the electrolyte (or current density required per unit of surface area)
and the total thickness of the unit cell for the three cases under con-
sideration. The calculations are based on a common volumetric power
density of 100 W/L, a 2 V emf and a 90% energy efficiency for the
overall electrochemical process. The approximate surface areas required per kilowatt hour for the three ranges of thickness have been
added for the reader's information. For polymer electrolytes, the
surface capacitance used is arbitrarily taken as 7 C/cm^2, a figure
which eloquently illustrates how a high surface-to-thickness ratio,

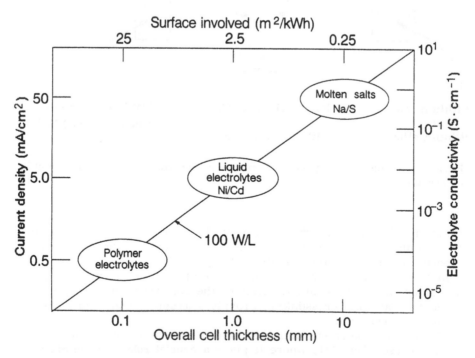

Fig. 5   Characteristics of electrochemical cells corresponding to
three battery technologies at power densities of >100 W/L.

typical of plastic film, can more than adequately compensate for a moderate conductivity.

C. Polymer Electrolyte Properties Required for Rechargeable Lithium Batteries

The lifetime (>2 years) and number of deep discharge/charge cycles demanded of a rechargeable battery are stringent constraints as far as the individual components are concerned. An electric vehicle, for example, will undergo over 800 deep cycles, whereas small consumer batteries will be subjected to fewer, but nevertheless several hundred, cycles. Their service conditions therefore impose severe mechanical, thermal, and electrochemical constraints on batteries, which indisputably represent an extreme case compared with most electrochemical applications.

Among the many components of a battery, it is the electrolyte that is exposed to the most stringent constraints, not only because it is omnipresent but also because of its triple function: Li^+ ion carrier, mechanical separator, and binding agent for the various electrode materials, as mentioned above. With respect to the mechanical and thermal constraints, cell cycling causes local variations in volume at both electrodes: $-dV$ at the lithium electrode and $+dV$ at the positive electrode during discharge. These variations, reversed on charge, are important at the positive electrode, but can be attenuated by choosing ion-insertion electrodes, for example $dV \sim 10\%$ for TiS_2. Displacement of the plastic separator during cycling allows more or less complete compensation for such localized volume changes. Other constraints of thermal origin arise if the battery has to operate at high temperatures, with differential thermal expansion taking place among the components; this, too, has to be absorbed by the polymer electrolyte. There are convincing examples, however, suggesting that polymer electrolytes can be adapted in such a way as to survive these constraints. A good example is automobile tires, which are made of carbon-black-charged rubber, and which demonstrate daily how polymers can survive harsh environments.

The greatest challenge for polymer electrolytes, however, comes from the electrochemical constraints. The minimum electrical conductivity requirements were discussed in the previous section, while the next sections will deal with the need to optimize the cationic nature of ion migration. The most important issue to be dealt with at the present time is the electrolyte's electrochemical compatibility with the other battery components: electrodes, collectors and insulators. From this point of view, batteries definitely represent an extreme case among most electrochemical devices. Totally reversible chemical or electrochemical reactions with almost 100% conversion are rare; in addition, lithium batteries, as shown in Figure 4, suppose that all secondary, or parallel, reactions are entirely suppressed.

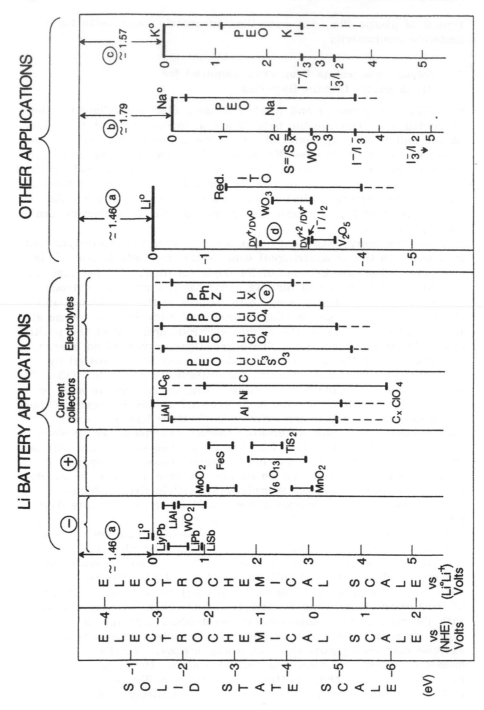

Yet this is only possible if quasi-perfect compatibility can be maintained between the battery components under storage, during discharge and, above all, during recharge, where a maximum stability window is required of the electrolyte.

Figure 6 presents approximate values for the redox properties of the main components of a battery, which gives an indication of the constraints they impose on the electrolyte. In fact, the latter is expected to have a wide enough stability window to be able to resist oxidizing couples in the form of positive electrode materials and reducing agents such as lithium. As can be seen from these few examples, the electrolyte, a complex formed from a solvating polymer and a dissociable salt, can have limited redox stability because of the characteristics of one or other of its constituents. Other materials located in specific parts of the battery can have a more limited redox range; such is the case of the current collectors, for instance, which are made of metal or carbon black.

It is obvious from the foregoing list of constraints, that the selection of a polymer electrolyte depends on the intended application. For any application, the total electrical conductivity is but a first test to be used in a selection process. The complete process can be summarized as follows:

Determine the total conductivity and the phase diagram
Determine the nature of the ion-carrier, such as $t^+_{cat} = 1$
Check for the possible role of impurities (such as protons,
    hydroxyl groups, or residual solvents)
Determine the redox domain in terms of the intended application,
    for example, cyclic voltammetry
Confirm the electrolyte in complete cells during intensive cycling
    or utilization tests

Fig. 6   Comparative redox scales for several components of electrochemical devices based on polymer electrolytes. (a), (b), and (c) 0 voltages of Li^o/Li^+, Na^o/Na^+, and K^o/K^+ scales are approximations based on aqueous system values, assuming similar hydration energies for water and polyether solvents. (From Ref. 23.) (d) Polydiviologen $\times$ $C1O_4^-$ couples. (From Ref. 24.) (e) Preliminary results. (From Ref. 25.)

IV. DEVELOPMENTS IN LITHIUM-ION-CONDUCTING POLYMERS

A. Properties of PEO Complexes

PEO-Salt Complex Preparation

Wright [5,6] has prepared PEO-alkali metal salt complexes using a solvent common to all the constituents, and this procedure is still widely used with solvents of the type CH_3CN and CH_3OH, despite the fact that in the long term a fusion-based (polymer-salt) process should really be more suitable for the preparation of plastic electrolytes.

The various factors governing the formation of solvating polymer-alkali metal salt complexes have been determined by Chabagno [14] and Shriver [26]. As a rule, the competing phenomena are, on the one hand, the solvating power of the polymer, which promotes salt dissolution, and, on the other hand, the high lattice energy or cohesive energy density values of the original products, which play the opposite role. This balance is expressed by the following reaction:

$$<M^+ \ X^-> + nPEO = PEO_n \ Li^+X^- \text{ (solid solution)} \tag{2}$$

The determining influence of the lattice energy has been clearly established by Shriver, who pointed out that in the case of salts with very high energies, complexes are not usually formed. The acceptable thresholds are nevertheless dependent on the cation used, since the latter also conditions the solvation energy for the complex. For example, a high threshold of -880 kJ/mole is acceptable for lithium salts because this cation has a strong polarizing power with respect to the PEO. Dissolution of the salt to form a complex is governed by Lewis acid-base interactions, which are similar to those of polar liquid organic solvents.

Thus the notions of donor numbers and atom hardness/softness are brought into play in cation solvation by the heteroatoms of the polymer, as is the case with oxygen of the ether functions, which carry donor electrons and form a strong Lewis base, DN $\sim$ 20 [14]. In aprotic media like these, anion solvation seems to play only a minor role, except when ion pairs are formed [27]. However, cation solvation by polymers also involves macromolecule-related considerations, including the density and distribution of the donor heteroatoms (CH_2—CH_2O configuration) and the flexibility of the EO chain, which allows intermolecular conformations that enhance solvation. These factors, as mentioned in section III, part A, have an important part to play in the formation of polymer-salt coordination complexes. PEO-metal salt complexes are now formed from a much wider range of products than the alkali metals and anions originally considered, such as I⁻, SCN⁻, ClO⁻₄ and CF₃SO⁻₃. Today, complexes involving alkaline earth salts and transition metals are known [14,28–30], substantially as a result of recent research on batteries [31–33].

Ionic Conduction Mechanisms and Role of T_g

Armand has explained the ionic conduction observed for PEO-alkali metal complexes by means of a model based first on the helical channel structure frequently associated with polymers and, second, on the conduction mechanisms in crystalline solids [10]. The basic crystalline complex has a PEO₄-Li stoichiometry at low temperature and presents a defective form at high temperature as shown in Fig. 7.

Specific cation conduction by a hopping mechanism within the helical solvating structure was postulated. The interest of this model at the time was its capability of explaining the maximum in the conductivity observed at an O/Li ratio close to 8/1, which corresponds to an alternation between vacant and non-vacant sites considered favorable for solid-state conductivity.

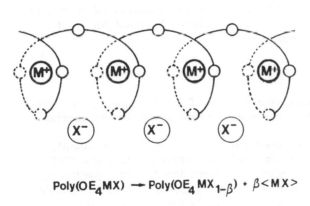

Poly(OE₄MX) ⟶ Poly(OE₄MX₁₋β) + β<MX>

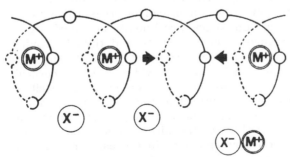

Fig. 7   Crystalline model of ionic conduction.

The model as such is no longer used to explain the conduction of
PEO-salt complexes, since the existence of amorphous phases and their
role in the conductivity of these electrolytes came to be known. NMR
studies of Li, F, and H have shown indisputably that the ions move
within the elastomeric phases, even when the latter coexist with crystalline phases of PEO or PEO_n-MX complexes [34]. Various studies have
been reported since then, confirming the preponderant influence of the
amorphous phases on the transport properties; for example:

Effect of ion-pair formation on conductivity [27]
Effect of electrolyte quenching on conductivity [35]
Conductivity versus temperature with respect to the phase
      diagrams [36,37].

The general conclusion to be drawn from these different studies is
that the ion transport properties in polymer electrolytes are strongly
dependent on the freedom of movement for cooperative rearrangement
of polymer segments and on the associated rearrangement of polymer
segments and on the associated local free volume.

The expression for the conductivity based on the free-volume
theory, as expressed by a VTF equation, corresponds fairly well to
the plot of the conductivity vs. temperature observed for most
largely amorphous PEO complexes:

$$\sigma = A.T^{-1/2} \cdot \exp -E_a/K(t - T_o) \qquad \text{(VTF equation)} \qquad (3)$$

where E_a is a pseudo activation energy, T_o is the ideal glass transition temperature, and A is a constant proportional to the carrier
concentration.

Despite the fact that the conductivity vs. temperature results can
often be represented in the form of an Arrhenius plot, the VTF representation offers the advantages of emphasizing the role played by
glass transition in conductivity and of suggesting ways to optimize
the conductivity by decreasing T_o. T_o is related by a constant to
T_g, the glass transition temperature, which can be measured experimentally. Meanwhile, several approaches are known for modifying
the glass transition temperature of an electrolyte and thus improving
its conductivity. The nature and concentration of the salt, for example, have a considerable effect on T_g [38,39] while various chemical
and physical modifications to the solvating polymer are also possible.

Several other parameters strongly influence the conductivity of
PEO-based electrolytes. For example, the real degree of dissociation
of the dissolved salt and the nature of the different species formed
have an effect on A, the term corresponding to the carrier concentration. The co-existence of amorphous and crystalline phases also

affects the conductivity. These factors will be examined in greater detail in the sections that follow.

Phase Diagrams

Interpretation of the thermal and electric behavior of PEO(linear)-MX complexes has made great strides since the notion of phase diagrams was first introduced [36,40]. This approach offers an explanation for (a) the presence of compounds having well-defined stoichiometries, such as $PEO_n \cdot MX$ [26,41,42], (b) the shape of DSC curves, and (c) the presence of crystalline phases observed at different compositions [43,44]. Figure 8 shows the phase diagrams of the two polymer electrolytes found most frequently in battery applications. The presence of well-defined compounds $(PEO)_3 \cdot LiCF_3SO_3$, $(PEO)_3 \cdot LiClO_4$, and $(PEO)_6 \cdot LiClO_4$, and of eutectics, has been confirmed by optical microscope observations, conductivity measurements and X-ray diffraction. These phase diagrams offer an explanation, at least in qualitative terms, for some of the aspects of the dependence of conductivity on temperature and salt concentration. Such is the case of the discontinuities frequently observed on plots of log σ vs. $1/T^\circ K$, which correspond to the eutectic temperature of the complexes [14,37]. The coexistence of a nonconducting crystalline phase of variable proportion with the conducting amorphous phase results in increased tortuosity and resistance of the electrolyte. This effect is also clearly discernible at high salt concentrations with O/Li ratios close to 3/1 and at low temperatures, where a high portion of crystalline phases is present (Fig. 8).

However, there is no straightforward treatment for the relationship between the conductivity and the phase diagram as a function, for example, of the phase rule and volume fraction of the coexisting phases. There are apparently too many factors simultaneously involved, such as the change in concentration and mobility of the carrier ions depending on the temperature or the O/Li ratio. The isotherms in Fig. 8 show that the conductivity reaches several maxima. The maximum value observed at low salt concentrations shows just how complex the competition between concentration and mobility is, even when the electrolyte is amorphous and monophasic.

Other phase diagrams involving PEO are known: with NaSCN [45], $LiAsF_6$, NaI, and LiI [46], all of which share a number of common general characteristics (well-defined compounds, presence of eutectics). Observable differences seem to be due, in part, to the role of the anion.

The use of phase diagrams leads to a better understanding of the behavior of polymer electrolytes and should be included in the characterization of any new solvating polymer-metal salt complex, except for entirely amorphous or reticulated polymers.

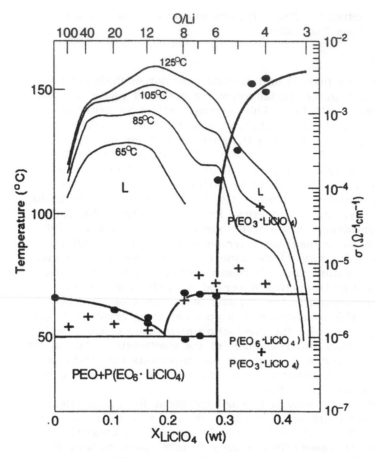

Fig. 8   Phase diagrams and conductivity isotherms vs. concentration for PEO-LiClO$_4$ and PEO-LiCF$_3$SO$_3$ complexes. (From Ref. 37.)

Cationic Transport Numbers

An electrochemical generator operating with reversible Li electrodes (Fig. 4) obviously has a lithium-ion-conducting electrolyte. The first crystalline model put forward by Armand to explain conductivity in polymer electrolytes (Fig. 7) suggested a specifically cationic conductivity [10]. It was only later, around 1982, that doubts began to be voiced as a result of voltammetry studies [13], impedance measurements [47,48], chronoamperometry [49], and salt and anion diffusion in PEO complexes [50]. Determination of the strong influence of elastomeric phases on conductivity and of the rapid diffusion

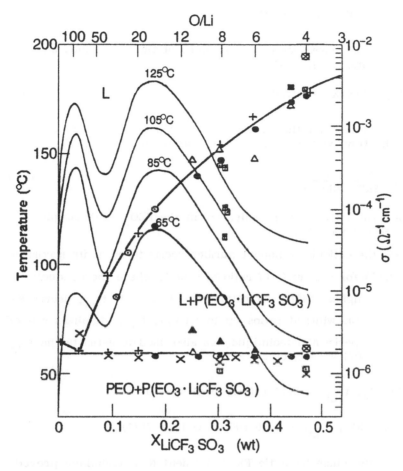

Fig. 8 (Continued)

of anions in these phases [41] suggests an analogy between polymer electrolytes and liquid organic electrolytes. In CH_3CN, DMSO or PC-$LiClO_4$ electrolytes, for example, the cationic transport number ranges from 0.2 to 0.4.

Systematic measurements have recently been made of the transport numbers in amorphous electrolytes PEO—LiI, PEO—$LiClO_4$ and PEO—$LiCF_3SO_3$.

The transport numbers of the PEO—LiI system have been determined [24] by parallel electrochemical measurements on concentration cells with junctions of the type:

$$Li^{\circ}/PEOm_1\!-\!LiX//PEOm_2\!-\!LiX/Li^{\circ} \tag{4}$$

where m_1 and m_2 correspond to two values of the O/Li ratio, and, also, on formation cells of the type:

$$Li^{\circ}/PEO_m\!-\!LiX/X^- \text{ electrode} \tag{5}$$

The emf expression for the first type of cell provides a means of deriving the transport numbers of the constituent, namely $t^c_{X^-}$

$$t^c_{X^-} = \frac{F}{2RT} \frac{dE}{d\ln a\pm} \tag{6}$$

since the mean activity of the salt, $a\pm$, in the $PEO_m\!-\!LiX$ complex can be obtained from Eq. (5).

At 90°C, the cationic transport number seems to be at its maximum ($t^c_{Li^+} = 0.45$) for a mean O/Li ratio of $\sim 30/1$, where the conductivity is also at its highest [46]. At ratios less than 10/1, $t^c_{Li^+}$ decreases rapidly to <0.10, while at ratios up to $\sim 120/1$, $t^c_{Li^+}$ stabilizes around 0.30. This experimental technique was also used to estimate the $t^c_{Li^+}$ values of the following electrolytes:

$$PEO\!-\!LiClO_4, \quad t^c_{Li^+} \sim 0.21 \; (90°C, \; O/Li \sim 8/1)$$

$$PEO\!-\!LiCF_3SO_3, \quad t^c_{Li^+} \sim 0.75 \; (90°C, \; O/Li \sim 50/1)$$

However, the Pulse Magnetic Field Gradient NMR technique proved most efficient for characterizing the transport numbers of these two electrolytes commonly used in batteries. This technique gives access to macroscopic diffusion coefficients and, indirectly, to the transport numbers.

In the case of the PEO–LiClO$_4$ system at 112°C, t^+ varies between 0.29, 0.18, and 0.28 for O/Li ratios of 6/1, 8/1, and 20/1 [44]. For the PEO–LiCF$_3$SO$_3$ system and an O/Li ratio of 8/1, t^+ varies regularly from 0.34 to 0.41 between 155°C and 175°C [51].

In other work on compounds similar to lithium triflate [52], PEO–$LiC_xF_{2x+1}SO_3$ also shows values of t^+ between 0.21 and 0.41 for O/Li ratios of 20/1 and 6/1 and values of x of 6 and 8. The general conclusion drawn from these studies is that the cationic transport number is lower than its anionic counterpart, thus confirming the

analogy with aprotic liquid organic electrolytes. This similarity
suggests that the salt in polymer electrolytes is not strongly dissociated ($\varepsilon \sim 5$ for PEO) and is present in the form of ion pairs,
multiplets, or ion agglomerates or clusters [53]. The conductivity
would therefore appear to be a result of the dissociation of these
ion aggregates to produce more mobile cations and anions. The
cause for the lower mobility and cationic transport numbers could
be strong cation-solvating chain interactions or statistical considerations but it remains unclear. However, with rising temperatures, an
increase in t^+ is observed as well as a tendency toward equal participation of anion and cation: $t^+ = t^- = 0.5$ [24,52].

The scarce information available [28,47] on PEO complexes with
multivalent cations suggests nevertheless that conductivity is mostly
anionic in character.

Advantages and Limitations

PEO-based polymer electrolytes offer very broad redox stability
windows, especially when salts like $LiClO_4$ and $LiCF_3SO_3$ are used.
As seen in Fig. 6, these materials show such good compatibility with
oxidizing insertion materials such as V_6O_{13} that they can be re-
charged at significant rates, that is, at voltages 200–500 mV higher
than predicted by thermodynamic values. Even in the case of oxida-
tion-stable anions like triflate, it is possible that the upper oxidation
limit, ~ 4.3 V, is still related to oxidation of the anion rather than
of the polymer [47,54,55]. This limit is also found to be very
temperature-dependent.

The compatibility of PEO complexes with lithium under redeposition
conditions is well established [47,56,57], but it is not the result of
the thermodynamic stability of the electrolyte with respect to lithium.
It is due rather to the formation of passivating films at the lithium
surface [46,58,59]. The stability of either functions, or their metastability as favored by the solid state, makes PEO complexes excellent
electrolytes for rechargeable lithium batteries. In addition, the
solvating power and flexibility of the CH_2—CH_2—O segments gives
these polymers an unsurpassed conductivity in high-temperature applications, as seen in Fig. 9. However, the strong crystallinity of
linear PEO, $\sim 80\%$, and the tendency of PEO-salt complexes to form
stoichiometric crystalline compounds and eutectics at temperatures
generally above 40°C correspond to a drastic reduction in conductivity
below 60–40°C. The almost total disappearance of the conducting
amorphous phases at 23°C (Fig. 8) results in a conductivity usually
lower than 10^{-7} S/cm. It is clear that PEO-based electrolytes offer
little interest for rechargeable batteries operating at room temperature
but they are still attractive for high-temperature applications.

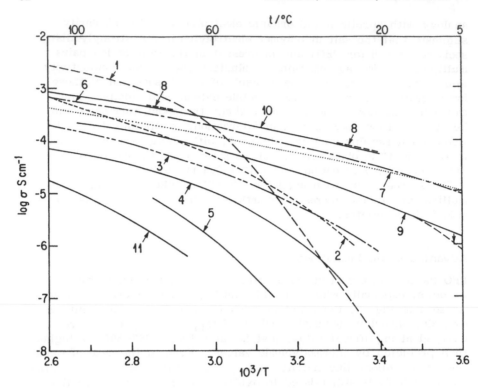

Fig. 9   Conductivity vs. 1/T°K for polymer-salt complexes:

1. PEO, LiClO$_4$ (12/1). (From Ref. 14.)
2. PEO, gamma-radiation crosslinked, LiClO$_4$ (1/8). (From Ref. 93.)
3. PPO, LiCF$_3$SO$_3$ (9/1). (From Ref. 10.)
4. Polyethylene adipate, LiCF$_3$SO$_3$ (4/1). (From Ref. 21.)
5. Polyethylene succinate, LiBØ$_4$ (6/1). (From Ref. 20).
6. Polyphosphazene ME7P, LiCF$_3$SO$_3$ (16/1). (From Ref. 33.)
7. Polyphosphazene crosslinked 1X MP, LiCF$_3$SO3 (32/1). (From Ref. 94.)
8. Polysiloxane PMM57 linear, LiClO$_4$ (25/1). (From Ref. 84.)
9. Triol type PEO crosslinked with difunctional urethane LiClO$_4$ (50/1). (From Ref. 99.)
10. PEO-PPO-PEO block copolymer crosslinked with trifunctional urethane, LiClO$_4$ (20/1). (From Ref. 77.)
11. Poly(N-methyl arizidine), LIClO$_4$ 8/1. (From Ref. 8.)

B. Improvement in Conductivity Complex by
    Polymer Modification

Studies of PEO-LiX complexes demonstrate very clearly that ion conductivity is present in the amorphous phase, not in the crystalline
phases. On the one hand, conductivity in the electrolyte seems to
depend on the mobility of the polymer chain segments, as characterized by T_g. On the other hand, the ion carrier concentration depends not only on the salt concentration but also on its degree of
dissociation, which is related to the dielectric constant of the polymer. Optimization of Li-conducting polymer electrolytes for room-
temperature applications can, therefore, be achieved by modifying the
solvating polymers involved in the electrolyte.

Considering the outstanding redox stability windows of PEO complexes and the high solvating power of the basic monomeric units
CH_2—CH_2—O (EO), it seemed that the first modification step would
be to try to improve this electrolyte family by changing the properties of the ethylene oxide chain. However, the changes envisaged
should not jeopardize the electrochemical stability of the ether
functions nor the solvating power of the EO segments.

A rough classification of the modifications proposed for polyethers
is presented below.

Crystallinity Reduction

The stereoregularity of the linear PEO chain means it has a high
degree of crystallinity, $\sim 80\%$. The melting point for high molecular
weights ($>10^6$) is $\sim 65°C$, T_g is $-60°C$, and the density is 1.15.
The highly crystalline nature of pure or complexed PEO can be observed by the naked eye in the form of spherulites, which sometimes
reach up to 1–3 mm on thin films (see Ref. 37 for instance). Crystallinity can be reduced in various ways: chemically, structurally,
and by a combination of these methods, as explained below.

Chemical Approach. It is possible to create disorder in the EO
chain by introducing one or several monomers or different sequences
of the ethylene oxide. These defects can be solvating or not, and
can be distributed statistically, alternately, or sequentially. This is
the very basis of copolymers, examples of which are given in the
literature [60–63], and schematically illustrated in Fig. 10. The
positive effects of reduced crystallinity usually become visible at
temperatures below 60°C, and published data tend to give room-
temperature conductivities of the order of 1–5×10^{-5} compared to
10^{-7}–10^{-8} S/cm for equivalent PEO complexes.

Structural Approach—Substituted Linear Structure. The first
approach put forward for reducing the crystallinity of PEO involved

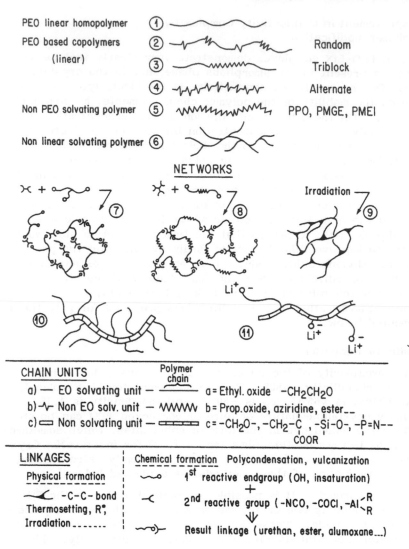

PEO linear homopolymer ①

PEO based copolymers ② Random
(linear) ③ Triblock
④ Alternate

Non PEO solvating polymer ⑤ PPO, PMGE, PMEI

Non linear solvating polymer ⑥

NETWORKS

⑦   ⑧   Irradiation ⑨

Li⁺

⑩   ⑪

CHAIN UNITS   Polymer chain
a) — EO solvating unit —   a = Ethyl. oxide $-CH_2CH_2O$
b) Non EO solv. unit —   b = Prop. oxide, aziridine, ester __
c) Non solvating unit —   c = $-CH_2O-$, $-CH_2-\overset{|}{\underset{|}{C}}-$, $-Si-O-$, $-\overset{|}{P}=N--$
                                                 COOR

LINKAGES | Chemical formation Polycondensation, vulcanization
Physical formation | 1ˢᵗ reactive endgroup (OH, insaturation)
                        | +
— -C-C- bond | 2ⁿᵈ reactive group (-NCO, -COCl, -Al$\langle$ᴿ_ᴿ)
Thermosetting, R˙; | ⇓
Irradiation ...... | Result linkage (urethan, ester, alumoxane ...)

Fig. 10   Examples of polymer derived from ethylene oxide (EO) configurations:

1. PEO = good solvation, high conductivity at >80°C, room-temperature crystallization
2. Crystallinity reduction due to chemical disturbance
3. Crystallinity reduction due to chemical disturbance
4. Crystallinity reduction due to chemical disturbance
5. Less crystallizable monomer unit resulting in amorphous (but less solvating) polymer at room temperature

an attempt to prevent regular organization of the linear chain by placing substitutes in the EO monomeric unit. Examples include non-stereoregular homopolymers such as polypropylene oxide (PPO) or polymethylglycidylether (PMGE), which give amorphous complexes [14] but have a poor conductivity compared with that of PEO (see Fig. 9). This is probably a result of their reduced solvating power or steric effects, which curtail the ion mobility.

Structural Approach—Nonlinear Structure.   PEO crystallinity can be reduced if the polymer has sidechains or if it corresponds to a starlike or comblike structure. In these cases, the sidechains remain free and the polymer is theoretically soluble (nonreticulated). A few examples can be found in the literature [64,65], while Fig. 10 presents an illustration of some conceptual possibilities.

Structural Approach—Reticulated Structure.   Chemical or physical reticulation represents a powerful means of preventing the crystallization of PEO chains. From a practical point of view, it gives the network elastomeric properties which oppose the tendency of the electrolyte to creepage, as is the case for polymers with a linear chain. Electrolytes consisting of polyether (PEO or PPO) networks and alkali metal salts have been studied systematically by Cheradame et al. [66–70] and by Watanabe et al. [71–74]. The linkage agents employed are usually multifunctional isocyanates and substituted silanes. In these studies, the starting polyethers (PEO, PPO) have a low molecular weight ($\sim$400–10,000) and are present as homogeneous, block-type diols, or triols (see Fig. 10).

A systematic characterization of chemically reticulated electrolytes was undertaken concurrently with the work on linear PEO-LiX complexes. Measurements of the conductivity [66,68], viscoelasticity [75] and diffusion of lithium by NMR [76] converge toward a VLF type, free-volume model and emphasize the importance of the glass temperature. LeNest [77] associates the displacement of the ions with movements of the chain segments and uses the reduced-temperature

---

Fig. 10   (Continued)

   6.   Structural disturbance: reduced crystallinity
   7.   PEO-based triol + diisocyanate polyurethane network
   8.   Triblock OE.OP.OE diol + triisocyanate (or + AlR3) poly-
      urethane (or alumoxane network)
   9.   Physical crosslinking (gamma-radiation, free radicals)
 10.   Nonsolvating (low T_g ?) backbone ($-SiO-$, $-P=N$, butadiene)
      with solvating sidechains (comb-like polymers)
 11.   Autoionizable polymers: $-O^-$ strong anionic groups; Li^+
      associated lithium ion

concept (T − T$_g$) to establish a relationship between the conductivity and concentration at constant free-volume. This author also explains the influence of the salt concentration of T$_g$ by an effect of ionic reticulation nodes of the solvating chains by the lithium salt. Two conductivity curves of polyether networks with LiClO$_4$ are given in Fig. 9. The transport numbers of several such systems have been measured [65,69], mainly as a function of the length of the solvating chain and the concentration and nature of the salt. Anionic conductivity tends to prevail and increases with the chain length: t$^+$ = 0.05 for 45 EO monomeric units, compared to t$^+$ = 0.23 for 23 EO units (O/Li = 20/1). The increase in the LiClO$_4$ concentration also reduces the cationic transport number; for example, for chains comprising 23 EO monomeric units, t$^+$ varies from 0.23 to 0.37 when the O/Li ratio is changed from 8/1 to 60/1.

Other chemically prepared polyethers have been proposed [61], sometimes the purpose of reducing the volumetric fraction of the reticulation nodes to improve the conductivity [78,79].

Figure 10 presents an example of the reticulation of linear PEO by physical means. In Ref. 93, gamma irradiation is performed at high temperature in order to fix the amorphous state, which favors conductivity at low temperature (see Fig. 9).

Mixed Approach. Several examples have been reported showing the advantages of combining the chemical and the physical approaches described above in order to minimize electrolyte crystallinity without losing too many of its elastomeric properties [60,62,69,80]. A typical case is polyurethane networks formed from block copolymer diols (PEO-PPO-PEO), whose conductivity is shown in Fig. 9.

Reduction of T$_g$

Chemical Approach. The relation established between conductivity and T$_g$ suggests that research would be worthwhile now on solvating polymers with a T$_g$ lower than that of linear PEO, which shows a temperature of −60°C for the amorphous part. Both polyphosphazenes and polysiloxanes (T$_g$ ∿ −123°C) have chains that offer very high flexibility, allowing cooperative polymer conformation fluctuations as needed for cation solvation and mobility. However, these chains do not have the required solvating power to form conducting complexes with Li salts, and EO monomeric units (CH$_2$—CH$_2$—O) are therefore added to make up for this. The solution usually proposed and originally reported in the literature by Shriver, Allcock, et al. [80, 81] is to add polyether sidechains to the polyphosphazene polymer backbone (Fig. 10). The influence of the optimum length of polyether chains on the electrical conductivity has been studied by Blonsky, who also investigated the reticulation effects [80]. Figure 9 provides an illustration of the conductivity obtained with these comb-like polymers whose transport numbers t$^+$ seem to vary between 0.2 and 0.4.

Other comb-like structures possessing a flexible chain have been proposed for polysiloxanes [83,84], which yield room-temperature conductivities of the order of 5×10^{-5} S/cm with optimum sidechains made up of seven EO monomeric units (see Fig. 9). More recently, it has been suggested [85] that a chain derived from polybutadiene be used with substituted alkoxy sidegroups. However, the conductivity values for short chains (two EO monomeric units) are still low, namely $\sim 4 \times 10^{-8}$ S/cm with NaI.

Physical Approach with Plasticizers. Widespread use is now made of plasticizers which, among other purposes, help to reduce the T_g of industrial polymers. This approach has been seriously considered as a means of improving the conductivity of polymer electrolytes [86-88]. The use of polyethylene glycols, for example, introduces OH end groups, which not only have an effect on cation solvation but, even more important, are capable of reacting with the lithium negative electrode. If the OH is replaced by methoxylated groups, the lithium stability seems to be enhanced, while conductivities of $\sim 10^{-4}$ S/cm are obtained at 40°C [86]. Adding liquid products to the polyethers bears similarity to networks swollen with liquid solvents [7]. In such a situation, the electrolyte properties are often governed by the liquid additive as far as the electrolyte's compatibility with the electrodes is concerned: namely, lithium passivation and oxidation or coinsertion at the positive electrode.

Increased Solvating Capability and Polymer Polarity

The solvating power of the polymer chain unit (CH_2—CH_2—O) and the degree of dissociation of the dissolved salt can be increased by replacing the oxygen by other donor heteroatoms such as nitrogen or sulfur. At the limit, other organic groups can be used instead of the ether function.

Polyimines. Use of nitrogen on a solvating monomeric unit of the type (CH_2—CH_2—NR) favors cation coordination because of the high donor number of amine functions: DN = 61 for triethyl amine, as compared to DN = 20 for THF or ~ 22 for PEO [14]. Moreover, the strong similarity between solvating polymers and crown ethers seems to confirm this, since fully or partially nitrogenized cages are used as complexing agents for alkali metal salts [89,90]. The use of (CH_2—CH_2—NR) monomeric units in lithium-conducting polymer electrolytes has been envisaged [8,17,18], but the conductivity values reported so far are disappointing: $\sim 10^{-7}$ S/cm for (CH_2—CH_2—NH)$_6$ complexes with sodium salts. If the hydrogen atom present on the nitrogen is replaced by a methyl, the electrochemical compatibility of these electrolytes with lithium electrodes should be improved, but the poly(N-methyl aziridine) with $LiClO_4$ (N/Li = 8/1), as reported in Fig. 9, has a very low conductivity and the synthesis and characterization of polyimines remain difficult.

Poly(alkylene) Sulfides.   These sulfides, including the ES monomeric unit (CH_2-CH_2-S), do not appear to form complexes with
lithium salts [91]. Shriver et al. attribute these results to differences in the conformation of the PES and PEO chains due to the
C—O and C—S bonds (1.43 A and 1.82 A, respectively), and to
the van der Waals radii of oxygen and sulfur (1.52 A and 1.85 A,
respectively). These differences are the reason for the sulfur
atoms of the PES chain not having such a marked "endo" character
as those of PEO when they are in a helical arrangement. In other
words, the sulfur atoms are not mostly facing the center of the
helix and therefore do not have the same intramolecular solvating
power as their polyether or crown ether counterparts.

Polyesters and Other Solvating Polymers.   The formation of polymer-
salt complexes has been demonstrated using poly(ethylene succinate)
and poly(ethylene adipate) with lithium salts [92–94]. These complexes have a strong concentration of polar groups, but their T_g
values are equivalent to or higher than that of PEO. Crystalline
complexes exist but, as with PEO, the conductivity is limited to the
amorphous phase and is significantly lower than the value observed
for PEO-containing electrolytes (see Fig. 9). The dielectric constant,
ε, which is probably higher for these polar polymers than for PEO,
should have a favorable effect on both the degree of salt dissociation
and the conductivity. However, this is not observed because it
seems to be masked by the effect of high T_g values on the conductivity as well as by the high cohesive energy of these polymers.

C.   Improving the Conductivity by Changing
the Salt and Its Concentration

Effect of Lithium Salt

The nature and concentration of the salt used for charging a polymer
have a complex influence on the electrolyte characteristics, as can
be seen from the phase diagrams and conductivity isothers of the
PEO-$LiCF_3SO_3$ and PEO-$LiClO_4$ systems in Fig. 8. The influence of
salt is easier to study in reticulated electrolytes because large crystalline phases are prevented [68,77]. It is important to bear the
following points in mind when studying the major effects of salt:

1.   The formation of the complex is largely due to the competition
between the lattice energy of the salt and the energy associated with cation solvation (see Eq. (2)). In a first approximation, the formation of the electrolyte calls for a polarizing
cation such as lithium (high solvation energy) associated with
a large anion and a delocalized charge, so that the lattice
energy of the salt can be reduced. The nature of this anion
also has a major impact on the characteristics of the phase

diagrams and on the composition of the polymer-salt crystalline complexes [95], possibly in relation to their reticular energy. This can be appreciated by comparing the two phases in Fig. 8.

2. The influence of the salt on the conductivity of the amorphous electrolyte affects two variables in Eq. (3): the constant A, which is related to the number of charge carriers, and the temperature T_0, which governs their mobility. The number of carriers depends on the degree of salt dissociation, which is controlled mainly by the dielectric constant of the polymer. Nevertheless the "acidic" character of the anion used can also contribute to the dissociation of the ion pairs [32]. Increasing the salt concentration is tantamount to an ionic reticulation of the chains, which increases T_0 [77]. This results in a decrease in the electrolyte conductivity at strong concentrations. The symmetry of the anion or the flexibility of its chain can sometimes introduce a plasticizing effect which lowers T_g and thus enhances the conductivity [96].

Salt Characteristics

The different approaches to maximizing the conductivity of polymer complexes discussed in section IV, part B are based on modifications to the solvating polymers. Another means of improving the properties of electrolytes involves the salt used to form the complex. The effect of salt on the properties of electrolytes, just discussed, stresses the critical role of the anion:

1. It has to be large and delocalized for the complex to form.
2. It must produce phase diagrams that allow an amorphous phase to exist over a broad range of temperatures and salt concentrations.
3. It must offer the required symmetry or chain flexibility to produce a plasticizing effect on the electrolyte.
4. It must correspond to the salt of a strong acid to curtail ion-pair formation.

Moreover the salt concentration must be kept as low as possible to avoid reducing the ion or chain mobility by increasing the T_g values.

Table 1 lists most of the salts used to form lithium conducting electrolytes. When the electrolyte is to be used in a rechargeable battery, however, the number of available salts is far more limited. Armand et al. [31,96] have proposed various families of salts and have formulated criteria designed to reconcile the solubility, conductivity, and redox stability requirements for polymer electrolytes. The required redox stability window vs. the lithium electrode is ~ 3.5 V or more, and applies to the salt as much as to the polymer (see

Table 1 List of Salts Proposed for Polymer-Lithium Salt
Electrolytes

Common lithium salts:

$LiCl$, $LiBr$, LiI, $LiSCN$, $LiClO_4$, and $Li\ NO_3$

Lithium salts with large anions:

Monovalent anions:

$LiB\phi_4$, $Li(C_5H_{11}-C\equiv C)_4B$, $Li(C_4H_9-C\equiv C)_4B$,

$Li(C_6H_5-((CH_2)_3-C\equiv C)_4B$, $R-COO-CH_2-CH_2-SO_3Li$
(where R is a nonsolvating polymer chain)

Divalent anions:

$Li_2B_{10}Cl_{10}$, $Li_2B_{12}Cl_{12}$ and $Li_2B_{12}H_{12}$

Fluoride compounds:

Monovalent anions:

$LiCF_3SO_3$, $LiC_4F_9SO_3$, $LiC_6F_{13}SO_3$, $LiC_8F_{17}SO_3$, $LiCF_3CO_2$,
$LiN(CF_3CO_2)_2$, $LiN(CF_3SO_2)_2$, $LiN(CH_3SO_2)_2$, $LiAsF_6$,
$LiBF_4$

Divalent anions:

$LiOOC(CF_2)_3COOLi$, $LiSO_3(CF_2)_3SO_3Li$

---

section III, part C, and Fig. 6). So far, only $LiClO_4$ and $LiCF_3SO_3$
have been used in rechargeable-battery applications and, in fact,
constituted the only salts recognized as fulfilling the electrochemical
stability criteria and offering adequate conductivity. Recently, however, new salts have been successfully tested as a substitute for
$LiClO_4$ in high-power cells for traction application [107] (see also
Figs. 12 and 14) and for their plasticizing properties in ambient-
temperature batteries [97].

D. Autoionizable Polymer Electrolytes

Conductivity Criteria

Marked anion mobility in electrolytes made from a charged polymer-
salt complex has a negative effect on the energy efficiency of a battery because it results in local concentration gradients (see section

V, part D). Various attempts have been made to curtail such mobility by increasing the size of the anions [32,33], their chain length [32,52,65], or their electric charge [31–33]. The salts listed in Table 1 have been studied, but most results are disappointing: Either the anion still has a high mobility, independent of its chain length [52], or the electrical conductivity obtained is too low because the salt dissolution is insufficient [32,33,65].

The best way to immobilize the anion is to fix it on the polymer chain, as in the case of the polyelectrolytes mentioned in section II, part B. The basic criteria to be fulfilled by polyelectrolytes in order to constitute dry solid electrolytes that are good conductors are:

1. Simultaneous presence of ionic groups with a fixed anion, along with cation-solvating groups
2. Choice of dissociable ion pairs for enhanced ionization of the salt (anionic groups derived from strong acids)
3. Sufficiently large number of ionizable groups and solvating functions to ensure cation displacement and conductivity values of the order of 10^{-5} S/cm at room temperature
4. Suitability flexible solvating chains for cation solvation and migration at fairly low T_g
5. Chemical and electrochemical stability of the various elements of the electrolyte for the intended applications

Preliminary Results

It is only recently that data have begun to be published on autoionizable lithium conductors. LeNest [77] and Lévèque [65,128] suggest the synthesis of autoionizable networks that use the ether functions (EO) as solvating groups. The ionic groups proposed are of the type:

$$
\text{Phosphate} \qquad
\begin{array}{c}
\text{O} \\
\parallel \\
-\text{O}-\text{P}-\text{O}-\text{O} \\
\mid \\
\text{O} \\
\text{Li}
\end{array}
$$

Lewis acid-lithium alcoholate complexes (e.g., $\text{R}-\text{O}-\text{SbCl}_5^-$, Li^+)
Perfluoro-carboxylates (e.g., $-(\text{CF}_2)_n-\text{COOM}$)

Bannister [32] describes a somewhat different approach, which involves mixing a polyether (solvating part) with the lithium salts of anionic polymers. Providing the molecular weights are sufficiently high and the anions are derived from strongly acidic polymers, the

mixture with the PEO yields an entirely cationic conducting electrolyte. A comblike polyether characterized by a methacrylate chain
coreticulated with sodium methacrylate has also been proposed as a
specifically cationic polymer [98].

A specifically cationic conductivity can be obtained with these
electrolytes, but applications are limited, either because the conductivity is low or because some of the electrolytes are not electrochemically stable. The complexity of the chemical reactions involved,
together with the difficulty in matching the various factors needed
to reach the required conductivity, donor numbers, chain flexibility,
T_g, and cation mobility between sites, must nevertheless be assessed
in terms of the advantages to be gained (in terms of battery performances, for example). These polymer electrolytes seem to present
the ultimate as far as future generations go.

E. Conductivity Performances and General Trends

The formation of polymer-salt complexes sufficiently dissociated to
function as electrolytes is not unique to PEO, and other polymer
electrolytes have been identified. However, ten years' research
into lithium-conducting polymer electrolytes has led to the conclusion that the most promising complexes still owe their high solvating capability and conductivity to EO-based constituents.

Figure 9, a particularly good illustration of this point, shows that
linear PEO always yields the highest conductivity values at elevated
temperatures (e.g., 2×10^{-3} S/cm at 100°C), when the electrolyte
is completely or mostly amorphous. Different modifications to the
PEO have managed to improve the conductivity at lower temperatures
by preserving the amorphous state, with a possible threshold value
of 10^{-4} S/cm. Such modifications, however, have been made at the
expense of the conductivity values at higher temperatures. Although
this cannot be attributed to a simple volume effect—percentage of the
solvating EO elements vs. percentage of nonsolvating elements (polymer network nodes, backbones or backbone segments, comonomers,
etc.)—this factor probably plays an important role. However, other
parameters must also be taken into account, such as the degree of
percolation of the solvating domains in the electrolyte and the effect
of the macromolecular structure on the distribution of the conductivity
between the anions and the cations.

What emerges from the electrolyte studies conducted so far is the
importance of the amorphous state and, in particular, the still determining role of the EO chains on the cation solvation capability and
on the conductivity. The configuration, conformation, and flexibility
of the EO chain tend to favor a wound structure, a cage or a helix,
where the oxygens are inward-directed [5,15]. This feature is well
known in crown ethers, for PEO with a helical structure with a

repetitive 19.3 A unit pattern, and for crystalline sodium or potassium complexes with a more compact thread. This arrangement tends to enhance the intramolecular solvation of the cation which, from the entropy point of view, is favored. Contrary to crown ethers, however, the chain conformations are reversible and possibly adjustable in that they accommodate rotation around the C—O bonds. Such flexibility, in the amorphous state, allows competition between intramolecular solvation and partial intermolecular solvation of the cation by the solvating chains, creating a balance which plays a determining role in charge carrier displacement.

The significance of the mobility of the EO chain segments comes to the fore very clearly in the various studies of the conductivity, where the glass transition temperature has a determining role often expressed by a free-volume model. This model applied to polymer electrolytes has suggested, or provided an explanation for, the effect of various approaches used to optimize the conductivity, including the effect of concentration, modifications to the solvating polymer, and role of the anion.

There is still some doubt about the nature of the ionic species present in the electrolyte or contributing to the conductivity but, in view of the low dielectric constant ($\varepsilon \sim 5$) of the medium, the salt is probably present in the form of associated ions: ion pairs triplets, and quadruplets at low concentrations [27,37,53], or even in the form of aggregates or clusters at higher concentrations. The nature of the mobile species, for its part, is still a matter of controversy [24,53,77,99], possibly because it depends on the salts and polymers under study.

V. SOLID-STATE LITHIUM CELLS USING THIN-FILM POLYMER ELECTROLYTES

A. Main Goals and Major Challenges

Research on lithium-conducting polymer electrolytes began in the context of the 1970 world energy problems. For this reason, its main goal, from the outset, was the development of rechargeable high-energy density batteries, particularly for transportation purposes (electric vehicles). This orientation was to determine the technical objectives and the materials selection criteria:

150 Wh/kg and $\sim$100 W/kg
Over 700 deep-discharge cycles
Target cost $\sim$\$100/kWh

Polymer electrolytes obviously had to meet fundamental basic and technological requirements if these objectives and criteria were to be fulfilled:

Stability in contact with reactive electrode materials, lithium, and
    oxidizing materials
Resilience, in order to conserve the adhesive properties and main-
    tain the electrical contacts during sustained cycling
Potential for the development of new generations of conducting
    polymers at or below room temperature
Reliability of thin polymer electrolyte in large-surface multilayer
    devices that tolerate no electrical faults
Availability of low-cost technological processes for fabricating and
    assembling the battery components: lithium negative electrode,
    positive electrodes, electrolyte, and collectors
Availability of polymers, lithium salts, and electrode reactants that
    are neither dangerous, toxic, nor prohibitive in cost.

B. High-Temperature PEO-Based Cells

The initial focus of research on polymer electrolytes was on commercially available PEO. At high temperature, the conductivity of PEO
complexes seems adequate to meet the power density requirements of
electric vehicles. This approach was adopted both by the Anglo-
Danish group coordinated by Harwell Laboratories in the United
Kingdom and by the France-Canada group (ACEP project), and it
resulted in a first systematic study of the feasibility of polymer
electrolyte batteries [100,101]. The main materials explored at that
stage in the research and development work were V_6O_{13} and TiS_2,
as the positive electrode materials (insertion type), Li° and LiAl as
the negative electrode materials, and $LiCF_3SO_3$ and $LiClO_4$ as the
lithium salt. A schematic of the battery construction is presented
in Fig. 11.

The fabrication technology was based on solution casting of the
electrolyte and the positive electrode material. The data obtained
from this research served to assess the potential of such batteries
for electric vehicle applications [102–104], to draw up specifications,
and to establish preliminary costs. The energy and power density
requirements called for surfaces of the order of 15–25 m^2/kWh.

The crystallization of PEO complexes with $LiClO_4$ or $LiCF_3SO_3$ makes
them unsuitable for operation at temperatures below 60–70°C, as
expected from Fig. 8, whereas above 120–130°C, the redox stability
domain of the electrolyte decreases [55] and polymer degradation
occurs. Modified polyethers associated with new lithium salts have
been used recently [105] and yielded satisfactory performances from
the point of view of both cycling [106] and power density [106,107]
at around 60—80°C (see Figs. 12 and 14). Lower mean operating
temperatures for applications calling for power assure better stability
of the materials used and facilitate thermal management of large-
scale installations.

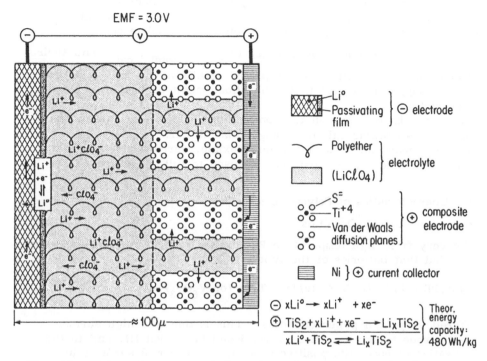

Fig. 11 Schematic representation of a polymer electrolyte Li battery with positive composite electrode.

C. Room-Temperature Cells

In 1983, research on new polyethers resulted in electrolytes sufficiently conducting at room temperature to be studied in complete cells [101]. The first results of initial cell testing demonstrated the feasibility of room-temperature rechargeable, all-solid batteries. These tests allowed the major characteristics of polymer electrolyte cells to be identified and the adaptability of polymer to different electrode materials to be assessed [108].

Concurrently, several laboratories began working on the development and characterization of room-temperature conducting polymer electrolytes. The basic approaches adopted to maintain the amorphous state of the EO chains were described in the previous section. Room-temperature conductivity values substantially higher

than those of PEO ($\sim 10^{-7}$–10^{-8} S/cm) have been achieved, as
illustrated in Fig. 9, but work remains to be done to establish the
characteristics of most of these electrolytes in complete cells,
especially their cationic conductivity ($\sigma_{cat.}^{+} = \sigma_{total}\, t_{cat.}^{+}$) and their
electrochemical stability. A few partial results have been published
[24,25,109], but suggest that applications might be limited to primary cells.

Constant improvement in the conductivity values and battery performance is drawing this technology towards markets where the
battery's characteristics—ultrathin, flexible, variable in shape, no
electrolyte leakage, low cost—can be used to full advantage.

D. Characteristics of Polymer Electrolyte

Power and Energy Densities

The very first high-temperature PEO-based configurations immediately
revealed that batteries of the type

$$Li^{\circ}/PEO-LiX, \ (O/Li \sim 10/1) \ / \ TiS_2 \ (or \ V_6O_{13}) \tag{7}$$

offered very high power densities. This is due to a high surface-
to-electrode-thickness ratio, as mentioned in section III, and to the
utilization of composite positive electrodes. The real surfaces involved at the positive electrode, >5 m^2/g, are large and the electrolyte acts as a binding agent which maintains the contacts and the
integrity of the active material during cycling.

Batteries studied under cycling conditions were not optimum configurations, in that they had low surface capacities (1–2 mAh/cm^2)
and the electrolyte layer was too thick, but they nevertheless
yielded high-temperature energy densities of around 100 Wh/kg [101,
110]. The corresponding gravimetric energy vs. power ratios are
presented in the form of Ragone plots in Fig. 12. Also in this figure are recent results for improved electrolyte cells operating at
lower temperature [107].

At lower temperatures, the transport properties of the polymer
electrolyte are rate-determining. Results published in 1985 mention
an energy content of between 90 and 150 Wh/L for discharge rates
of C/20 to C/50 (i.e., in 20–50 h) [101,108]. Figure 13 shows the
corresponding Ragone plots and compares these results to those obtained at high temperatures, on a volumetric basis. Results published as recently as 1987 [106] indicate higher discharge rates,
$\sim$C/5, for configurations similar to those used for earlier work,
namely $\sim$0.4 mAh/cm^2. This represents another major step towards
the goal of the ACEP project, namely, 70–300 Wh/L for rates of C/3
to C/200 [108], and it should yield higher power curves on the

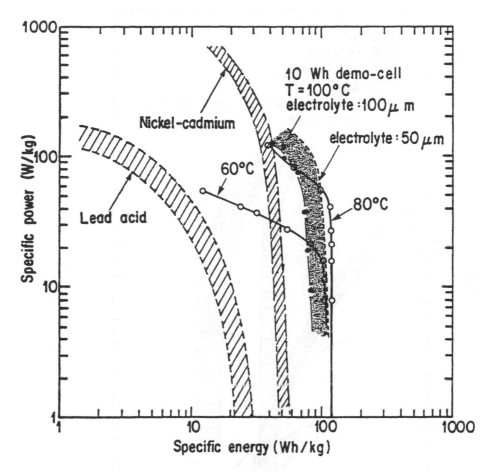

Fig. 12   Gravimetric Ragone plot for PEO cells at 100°C.   (From Refs. 110 and 107.)

Ragone plots, as represented approximately by the dotted lines in Fig. 13.

Cyclability and Conservation

Polymer electrolytes seem to favor the cyclability of lithium batteries [100,101,111,112], as can be seen in Fig. 14, which illustrates results obtained for various positive and negative electrode materials with different polymers and lithium salts.   Compared with liquid electrolytes, for example, where over 20 years' research has not produced more than a handful of formulations that allow lithium

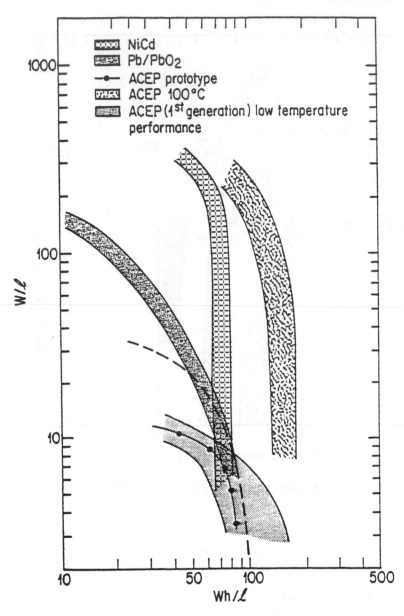

Fig. 13   Volumetric Ragone plot for room-temperature cells.
(From Refs. 111 and 107.)

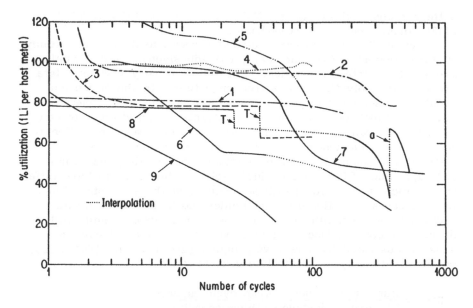

Fig. 14 % utilization for polymer electrolyte cells vs. cycle number (logarithmic scale). 100% utilization is defined as one-to-one ratio between inserting ion (Li^+ and metal atom of host structure (M)):

1. Li/modified polyether-$LiClO_4$/MoO_2, 4.4 mAh (3.9 cm^2), 60°C
2. Li/PEO 5M-LiX/VO_x, 2.9 mAh (3.9 cm^2), 80°C
3. Li/PEO 5M-$LiClO_4$/VO_x, 100 mAh (98 cm^2), 100−75°C
4. Li/modified polyether-$LiClO_4$/TiS_2, 640 mAh (2320 cm^2), 22°C
5. Li/PEO 5M-$LiCF_3SO_3$/V_6O_{13}, 0.75 cm^2) 135°C (From Harwell Laboratories, Ref. 84.)
6. Li-PEO 5M-$LiCF_3SO_3$/V_6O_{13} (From Harwell Laboratories, Ref. 83.)
7. LiAl/PEO 5M-$LiClO_4$/V_6O_{13}, 1.3 mAh (3.9 cm^2), 100°C
8. Li/PEO 5M-$LiClO_4$/TiS_2, 3.2 mAh (3.8 cm^2, 100−80°C, $a = I_D(0.5 \to 0.13$ mA/cm^2)
9. Li/polyphosphazene-LiX/MoO_2, 1.3 mAh (3.9 cm^2) 22°C (From Ref. 25.)

battery rechargeability (e.g., Moly Cell: Li°/PC-$LiAsF_6/MoS_2$)
[113], these results are remarkable.

The cyclability of polymer electrolytes seems to depend essentially
on three factors: (a) solid state, (b) adhesiveness and flexibility,
and (c) low surface capacities.   The solid, or very viscous, state
prevents the solvating chains from migrating with the cation and
avoids co-insertion and penetration of the solvent in the electrode
materials, which occurs with liquid solvents.   The absence of convection kinetically prevents the progression of local degradation reactions at the electrodes.   At the lithium electrode, where the
electrolyte is thermodynamically unstable, passivating films form
[58,59].   Such films exist in liquid solvents but seem to have different characteristics and growth mechanisms in the solid state [46].
In most cases, the film does not hinder the ionic and electronic
exchanges between lithium and the polymer electrolyte and good
lithium cycling efficiency is found over long periods of cycling.
This is illustrated by a relatively high figure of merit, FOM $\sim$ 120,
obtained with polymer electrolyte cells [112].

$$FOM = \frac{\text{number of cycles} \times \text{mean cycle capacity}}{\text{excess installed lithium capacity}}$$

With liquid electrolytes, FOM values in the range of 50 are frequently
observed.   This calls for a significant lithium excess in order to
achieve a long cycle life with liquid organic electrolytes.

As far as the mechanical properties are concerned, the electrolyte's
flexibility and adhesive properties ensure that electrical contacts are
maintained and prevent the lithium and positive electrode materials
from disintegrating or losing contact under cycling conditions.   The
full merits of the electrolyte's role as a binding agent come to the
fore in the cycling performances [106] and in *in situ* SEM observations of the behavior of different positive composite electrodes under
vacuum [114].   The polymer electrolyte technology, finally, is a surface technology requiring very low surface capacities, less than
4 mAh/cm^2, in fact.   Consequently, the current densities are always
low and, as a result, favor electrochemical redeposition of lithium
without dendrites.   The literature mentions the possibility of dendrite
formation [115], but such formation seems to be eliminated by an
appropriate control of the technological parameters [101,110].

The good storage capability of polymer electrolyte cells and the
absence of significant self-discharge distinguishes them from conventional lead-acid or nickel-cadmium batteries.   Self-discharge rates
of less than 0.1% a year at room temperature [108] and less than
2–4% over periods exceeding 4 months at 80°C have been reported
[107].   These data confirm the electrochemical stability of the
battery components, the absence of parallel secondary reactions, and
the very low value of the electronic conductivity, $<10^{-10}$ S/cm.

Scale-Up Reliability

One of the questions raised by the application of thin films (20–100 μ) in multilayer electrochemical devices concerns the reliability of polymer electrolytes with respect to electrical faults, which could result in short circuits between the electrodes or in uneven current distribution.

A partial response to this has been furnished by recently built laboratory prototypes using polymer electrolytes [110,111], as illustrated in Fig. 15. The 10-Wh prototype represents a high temperature configuration based on commercial PEO while the 1 Wh prototype is based on a low-temperature configuration and makes use of modified polyethers and electrolyte films some three to five times thinner than the usual PEO electrolytes. The surfaces involved are about 1000 times larger than those of the small cells (~ 4 cm^2) employed for electrochemical performance testing. No unfavorable scale effect has been observed in the performances of these prototypes, and the results, including cycling tests, confirm the extrapolation from smaller cells [111,112].

This verification of the effects of scale-up is still not sufficient for assessing the influence of cumulative volume variations or the thermal management of larger-scale installations, especially in special geometries such as the "jelly roll" assembly. However, preliminary calculations do not appear to indicate any insurmountable problems for thermal management [104].

Anion Transport Effect on Battery Performance

It was mentioned in section IV, part A, that anionic transport is present in most known polymer electrolytes. Anion participation in conductivity in batteries based on lithium reversible electrodes (see Fig. 4) results in concentration gradients close to the electrodes. The origin of this phenomenon and its effect on the energy efficiency of the electrochemical cell are illustrated in Fig. 16 for an ideal case in which only limitations related to the electrolyte are considered.

In practice, the severity of the limitation attributable to anionic transport is highly dependent on the working temperature and the cell configuration. At temperatures of $\sim 100°C$, this effect does not appear to dominate, and the cell performance for discharge rates of 2 C to C/2 is limited by the diffusion of Li$^+$ in the electrode material. However, at lower temperatures, the phenomenon is present and can limit the current at which a significant percentage of the cell material can be utilized [108].

E. Development Prospects

Energy storage in large polymer-electrolyte batteries for such applications as electric vehicles now seems closer to reality. Most of the

(a) Li^0/ mod. PEO, LiClO$_4$/ TiS$_2$, total area: 2400 cm^2
>100 deep DOD cycles, 80 Wh/l, (C/15 – C/30)

(b) Li^0/ POE 5M, LiClO$_4$/ Vanadium oxide, total area: 4700 cm^2
>100 deep DOD cycles, 100 Wh/kg, (3C – C/8)

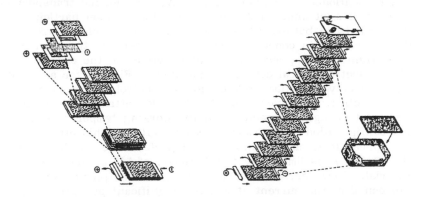

Fig. 15   10-Wh (a) and 1-Wh (b) IREQ prototypes. Cycling values correspond to full depth of discharge (DOD). (From Refs. 110–112.)

Role of t_{an}^- on local salt concentration gradients

a)   **Physical origin:**

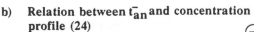

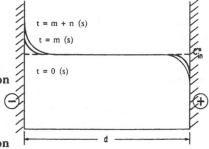

b)   **Relation between t_{an}^- and concentration profile (24)**

   for – **Homogeneous media**
   (amorphous electrolytes)
   – **D: Independant of concentration**
   – **Control by electrolyte**
   transport properties

$$C(x,t) = C_{in}^o - \frac{I\,t_{an}^-}{FD}\left(x - \frac{d}{2}\right)$$

Nonsymmetrical concentration profile if:

   – $\mu \propto (C)$
   (ex. $\mu \uparrow$ if $(C) \downarrow$ as $Tg \downarrow$)

   – **multiphased electrolyte**
   with crystallization/dissolution
   effects (e.g. $PEO_3\,LiClO_4$)

Global effect of t_{an}^- on cell efficiency

Effect on cell voltage of overvoltages due to electrolyte concentration profile:

$$E(V) = E_{th} - \eta_{diff} - \eta_\sigma$$

where: η_{diff} and η_σ correspond to the summation of all local diffusion and ohmic overvoltage contributions arising from changes in optimized initial salt concentration of electrolyte, ignoring overvoltage from other cell components.

Fig. 16   Relation between t^-, concentration profile, and overall cell efficiency.

basic problems that arose when these electrolytes first came into
use have been solved or are on their way to finding a solution. A
few aspects remain to be examined in closer detail, such as the performance over long cycling periods (>500 cycles) and the long-term
behavior of series-parallel arrangements. Furthermore, the thin-film
technology needs to be demonstrated for applications of the order of
1 kW to quell doubts about the assembly techniques, volume variation
control, and thermal management.

Recent improvements in the performance of room-temperature cells
and the ceaseless development of new, better-conducting polymer
electrolytes over a wider temperature range have narrowed the gap
between the technology and potential markets, where assets such as
thin-film construction, solid state, and flexible geometry prove
unique. Other characteristics, now firmly established, have already
laid the groundwork for short-term application of polymer electrolyte
batteries to microelectronics.

The only technological aspects that remain to be mastered for these
first applications to be successful on a large scale are the development of low-cost production and assembly processes for the cell components.

VI. OTHER POLYMER ELECTROLYTE APPLICATIONS

A. Energy-Related Applications

Most of the intrinsic advantages of polymer electrolytes that have come
to light in their use in lithium batteries can easily be transposed to
other electrochemical applications, such as energy generation or
storage. From the technological standpoint, for instance, the possibility of using the electrolyte in the form of adhesive thin films or
for fabricating multilayer all-solid devices has rekindled interest in
electrochemical devices based on surface technologies, such as photoelectrochemical and electrochromic cells. Polymer electrolytes lend
themselves particularly well to the modern trend toward all-solid,
miniaturized devices designed for easy-to-automate fabrication.

Good ambient-temperature conductivity ($\sim 10^{-4}$ S/cm) seems to be a
major prerequisite for such electrolyte applications. However, the
nature of the conductivity required for a given use must also be considered: cationic, anionic or mixed (ionic/electronic). Constraints
in terms of the redox stability of the electrolyte are usually not as
demanding as in lithium battery applications (see Fig. 6). As a rule,
it is oxidation stability that is more important, rather than stability
versus powerful reducing agents like lithium. A wider choice of
electrolytes can therefore be achieved if chemical modifications are
made to the polyethers or if new polymers bearing solvating functional
groups other than ethers can be found that need not be compatible with

lithium. The same remark applies to dissolved salts, which can also contain reducible anions such as polyiodides or polysulfides, or cations other than lithium, such as Na^+, K^+, and Mg^{2+}.

Photoelectrochemical Devices

The use of aprotic polymer electrolytes to make photovoltaic cells has been described by various authors [116–120]. A schematic view of the devices studied is presented in Fig. 17. These systems generally employ n-type semiconductors and transparent electrodes (indium-tin-oxide(ITO)-covered glass) and are characterized by complex redox couples of the type $I^-/I_3^-/I_2$ or $S^=/S_x^=/S_{x+1}^=$. Under illumination, the photons absorbed by the semiconductors produce electron-hole pairs which are separated in the space charge layer formed in the semiconductors in contact with the electrolyte. The holes (h) oxidize the complex anions at the semiconductor surface, whereas the corresponding reduction takes place at the transparent (ITO) counter

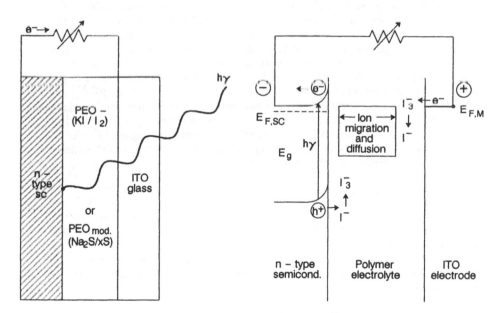

Fig. 17   Schematic view of a photoelectrochemical cell with energy level diagram where E_g is the semiconductor band gap, $E_{F,SC}$ is the Fermi level of the semiconductor and E_{FM} the Fermi level of the counterelectrode (ITO).

electrode. Rediffusion in the electrolyte of the species formed at the electrodes—I_3^- (I_2) or $S_{x+1}^=$ (s)—closes the electrochemical cycle whose global output is electrical work, similar to that of photovoltaic cells.

In addition to their interest as a solid-state technology, polymer electrolytes offer a means of freeing the semiconductor surface from photocorrosion, which represents one of the major problems associated with liquid electrolyte cells. They also allow the electrolyte thickness to be reduced, which means that the transparency is improved. The energy conversion efficiencies observed so far are low, $\sim$1–2%, since the technology is still in its infancy, but nevertheless the conversion appears to be extremely stable after a steady regime is attained.

One limitation that seems to hamper the performances at the present time is the low conductivity of the electrolytes used, based on PEO [116–118] or modified PEO [120]. Furthermore, the presence of mixed (anion + cation) conductivity in these electrolytes results in (a) local concentration effects, (b) a tendency for the oxidizable salts to become depleted close to the surface of the semiconductor, and (c) reducible-salt enrichment at the transparent electrode. This phenomenon is similar to the concentration effects observed in batteries (see section V, part D and Fig. 16). The outcome, under irradiation, is probably a profile of the salt concentration and, also, of the Fermi level of the redox couple throughout the electrolyte. This profile can extend deep into the polymer electrolyte, owing to the absence of convection, and cause local ohmic or diffusion overpotentials which undermine the overall energy efficiency of the device. However, for photochemical cells, the carrier species should be anionic, whereas for lithium batteries cationic transport is required. Known examples of entirely anionic polyethers exist which could be considered for this purpose [65].

Electrochromic Devices

So far, electrochromic devices have found few applications, mainly because of their performance limitations in terms of stability and reversibility, but also because alternative technologies, such as liquid crystals, offer strong competition. The devices envisaged with polymer electrolytes are represented by the following electrochemical reaction cell:

transparent	: optically	: polymer	: counter-
electrode	: active	: electrolyte	: electrode
	: redox	:	:
	: couple	:	:

In some cases, a third electrode is added to restore the capacity of the principal electrodes after several duty cycles. The optically active redox couple is either mineral, a thin film of WO_3 or V_2O_5, or organic and it can be activated by Li^+, Na^+, or other ions as well as by protons. Research to date has focused on protonic conductors [121,122] while modest progress has been made in work on aprotic polymers [24,123].

The use of adhesive polymer electrolytes significantly simplifies the fabrication of solid-state devices and could, theoretically at least, extend their range of application way beyond electrochromic display devices, such as in the direction of energy-efficient windows with electrochemically controlled reflection or absorption properties. The use of such devices in the windows of motor vehicles and buildings could significantly enhance the thermal management of incident solar energy [121].

B. Other Applications

Exploration of other possible applications for aprotic polymer electrolytes has only just begun. Apart from their technological merits, these polymer-salt complexes have the great advantage of offering a wide range of possible combinations between polymers and dissolved salts. First, there is the choice of various families of polymers, each with several possible modifications and, second, a multitude of soluble salts that could be used to produce conducting complexes [14,29,82]. In principle, use of different combinations of these parameters offers a means of controlling the nature of the ion-carrier species and adapting the electrolyte to specific applications, such as sensors. Typical examples include ion-specific electrodes, anion-responsive electrodes for measuring thermodynamic values [24] or specific-cation-containing electrolytes for potentiometric study of alloys [124]. Various types of ISFET or GASFET can also be envisaged. Other electrochemical applications based on specific-ion-bearing polymers are also being considered, such as lithium-free batteries that use magnesium or zinc [30], coulometers, or separation membranes based on anion carriers.

It should be noted, too, that the absence of vapor pressure with these electrolytes allows electrochemical reactions to take place even under high vacuum. For example, polypyrrole films used as transfer additives in photoelectrochemical devices have been electrochemically prepared *in situ* from polymer electrolytes during the fabrication of such devices under high vacuum [119]. High-vacuum electrochemistry with polymer electrolytes has also been used for the continuous observation of solid-state lithium battery electrodes inside a scanning electron microscope [114].

Exploration of mixed conductors (ionic + electronic) has just started [126], and application of electronically conductive complexes between polyethylene oxide and sodium polyiodide have been used to produce photovoltaic currents [127].

VII. CONCLUSION

The exciting prospects offered by polymer electrolytes today are due largely to a combination of technological factors and physicochemical properties which set them well apart from conventional liquid and solid electrolytes. As far as the technology is concerned, the plastic materials offer a means of chemically controlling the mechanical properties of the electrolyte (deformability, elasticity, and adhesion) and of fabricating it in the form of thin films. The physicochemical properties, in particular the solvating mechanisms and conductivity of polymer electrolytes, distinguish them from crystalline and glassy solids and strengthen their similarity to aprotic organic liquids. As with the latter, dissolution depends on cation solvation by coordination, although in the case of polymers it is their donor heteroatoms and the configuration of the macromolecular chain that ensure solvation, which as a result is multidentate and partially intramolecular. Ion conductivity is accordingly governed by the mobility of the solvating chains but, contrary to liquid electrolytes, the solvating part does not migrate with the ions. Another distinguishing characteristic of polymer electrolytes is their solid state, in which convection is suppressed, while the high viscosity modifies the transport properties and helps to confine degradation. This in turn contributes to a better chemical metastability of the "pseudo-liquid" electrolyte with respect to the electrodes.

Intensive research on lithium conducting polymer electrolytes for room-temperature applications had led to a far better understanding of these complexes and revealed the significant roles of the amorphous phases, as well as the usefulness of phase diagrams for electrolyte characterization. Various types of solvating polymer are known and many salts are now available for ensuring the conduction properties required for specific applications. So far, however, the polymers with the best properties with respect to solvation, conductivity, and redox stability are still based essentially on PEO or a high proportion of short chains composed of EO monomeric units. The superiority of the ethylene oxide chain resides in the solvating capability of the (CH_2-CH_2-O) configuration, adaptable conformation, and the chemical stability of the ether function.

The highly successful application of polymer electrolytes to lithium rechargeable batteries has brought out the qualities of these materials in surface devices, even in the presence of stringent mechanical and chemical constraints. Batteries using these electrolytes have

demonstrated high energy and power densities, remarkable storage
and cycling properties, easy-to-automate fabrication processes, and
a good form factor. If current research and development work continues to progress, we should soon see such batteries used for microelectronics applications and, in the long term, for mass electricity
storage which, among other functions, could herald large-scale use
of electric vehicles.

Wider use of polymer electrolytes is now expected, with applications foreseen in many areas, particularly those where the properties of this conducting material offer special advantages. Examples
include electrochemical surface techniques, miniature all-solid devices, systems calling for a chemical bonding between the electrolyte
and an electronic conductor (semiconductor or conjugated polymer),
and chemical or biological sensors, or other such applications where
controlling the nature of the polymer or the dissolved salt represents a means of performing specific functions.

ACKNOWLEDGMENTS

Société Nationale Elf-Aquitaine and Hydro-Québec are acknowledged
for authorizing the publication of this work. The authors also express their gratitude to Lesley Kelley-Régnier of Hydro-Québec for
translating the manuscript and to the word-processing personnel for
their devoted effort. A final word of thanks goes to Hydro-Québec
for bearing the costs associated with the preparation of this chapter.

REFERENCES

1. R. H. Baughman, J. L. Bredas, R. R. Chance, H. Eckhardt,
    R. L. Elsenbaumer, D. M. Ivory, G. C. Miller, A. F. Preiziosi,
    and L. W. Shacklette, *Conductive Polymers*, Plenum Publ. Corp.,
    New York, 1981, p. 137.
2. C. K. Chaing, S. C. Gau, C. R. Fincher, Y. W. Part, A. G.
    MacDiarmid, and A. J. Heeger, *Appl. Phys. Lett.*, 33:18 (1978).
3. L. W. Shacklette, J. E. Toth, N. S. Murthy, and R. H.
    Baughman, *J. Electrochem. Soc.*, 132:1529 (1985).
4. N. Mermilliod, J. Tanguy, and F. Petiot, *J. Electrochem. Soc.*,
    133:1073 (1986).
5. D. E. Fenton, J. M. Parker, and P. V. Wright, *Polymers*, 14:
    589 (1973).
6. P. V. Wright, *Br. Polym. J.*, 7:319 (1975).
7. G. Feuillade and P. Perche, *J. Appl. Electrochem.*, 5:63 (1975).
8. M. B. Armand, *Solid State Batteries*, C.A.C. Sequeria and
    A. Hooper, eds. NATO-ASI Series, 1985.

9. D. J. Barker, D. M. Berwis, R. H. Dahm, and L. R. S. Hory, *Electrochem. Acta, 23*:1107 (1978).

10. M. B. Armand, J. M. Chabagno, and M. J. Duclot, *Fast Ion Transport in Solids*, P. Vashishta, ed., North Holland, New York, 1979, p. 131.

11. M. B. Armand and M. Duclot, Eur. Pat. 0 013 199, Prior. FR 7832976.

12. K. M. Abraham, J. L. Goldman, and D. L. Natwig, *J. Electrochem. Soc., 129*:2404 (1982).

13. P. Rigaud, Doctoral Thesis, University of Grenoble, 1980.

14. J. M. Chabagno, Doctoral Thesis, University of Grenoble, 1980.

15. B. L. Papke, Doctoral Thesis, Northwestern University, 1982. Available from University Microfilms International 300 N Zeeb Road, Ann Arbor, Mich. 48106, U.S.A.

16. D. S. Newman, D. Hazlett, and K. F. Mucker, *Solid State Ionics, 3/4*:389 (1981).

17. C. K. Chaing, G. T. Davis, C. A. Harding, and T. Takahashi, *Macromolecules, 18*:825 (1985).

18. C. S. Harris, D. F. Shriver, and M. A. Ratner, *Macromolecules, 19*:987 (1986).

19. S. H. Clancy, Ph.D. Thesis, Northwestern University, 1985. Available from Univeristy Microfilms International 300 N. Zeeb Road, Ann Arbor, Mich. 48106, U.S.A.

20. R. Dupon, B. L. Papke, M. A. Ratner, and D. F. Shriver, *J. Electrochem. Soc., 131*:586 (1984).

21. R. D. Armstrong and M. D. Clarke, *Electrochem. Acta, 29*:443 (1984).

22. M. Watanabe, M. Togo, K. Sanui, N. Ogata, T. Kobayashi, and Z. Ohtaki, *Macromolecules, 17*:2908 (1984).

23. C. A. Hampel, *The Encyclopedia of Electrochemistry*, Reinhold Publ. Corp., New York, 1964, p. 415.

24. A. Bouridah, Doctoral Thesis, Engineering Dept., University of Grenoble, 1986.

25. M. Gauthier, *Rechargeable Lithium Polymer Batteries*, Third Int. Seminar on Li Battery Techn. and Applic., Florida, March, 1987. Available from Lithium Battery Conference Coordinator, Shawmco Inc., 14 Hawes Road, Sudbury, Masachusetts 01776.

26. D. F. Shriver, B. L. Papke, M. A. Ratner, R. Dupon, T. Wong, and M. Brodwin, *Solid State Ionics, 5*:83 (1981).

27. R. Dupon, B. L. Papke, M. A. Ratner, D. H. Whitmore, and D. F. Shriver, *J. Am. Chem. Soc., 104*:6247 (1982).

28. L. L. Yang, A. R. McGhie, and C. G. Farrington, *J. Electrochem. Soc., 133*:1380 (1986).

29. R. Hug and G. C. Farrington, Extended Abstracts of the Electrochemical Society Fall Meeting, San Diego, California, 1986, Abstract 765.

30. A. Patrick, M. Glasse, R. Latham, and R. Lindford, *Solid State Ionics, 18*:1063 (1986).

31. F. Cherkaoui, Doctoral Thesis, University of Grenoble, 1982.

32. D. J. Bannister, G. R. Davies, I. M. Ward, and J. E. McIntyre, *Polymer, 25*:1291 (1984).

33. J. S. Tonge, P. M. Blonsky, D. F. Shriver, H. R. Allcock, P. E. Austin, T. S. Neenan, and J. T. Sisko, Proc. of the Symp. on Lithium Batteries (San Diego 1986). Vol. 87-1, A. N. Dey, ed., The Electrochem. Soc., Pennington, New Jersey, 08534.

34. C. Berthier, W. Gorecki, M. Minier, M. B. Armand, J. M. Chabagno, and P. Rigaud, *Solid State Ionics, 11*:91 (1983).

35. D. F. Shriver, R. Dupon, and M. Stainer, *J. Power Sources, 9*:383 (1983).

36. P. R. Sorensen and T. Jacobsen, *Poly. Bull., 9*:47 (1983).

37. C. D. Robitaille and D. Fauteux, *J. Electrochem. Soc., 133*:315 (1986).

38. A. Eisenberg, K. Ovans, and H. N. Yoon, *Ions in Polymers,* ACS Symposium Series 187, A. Eisenberg, ed., Washington, D.C., 1980, p. 253.

39. H. Gheradame, A. Gandini, A. Killis, and J. F. LeNest, *J. Power Sources, 9*:389 (1983).

40. C. D. Robitaille and J. Prud'homme, 65th Canadian Chemical Congress, Toronto, 1982, Abstract NA-10. Available from the authors.

41. C. Berthier, Y. Chabre, W. Gorecki, P. Segrasan, and M. B. Armand, Proc. 160th Meeting A.E.C.S., Denver, 1981, p. 1495. Available from the authors.

42. M. Minier, C. Berthier, and W. Gorecki, *J. Phys., 45*:739 (1984).

43. J. E. Weston and B. C. H. Steele, *Solid State Ionics, 2*:347 (1981).

44. M. Gorecki, R. Andreani, C. Berthier, M. B. Armand, M. Mali, J. Roos, and D. Brinkmann, *Solid State Ionics, 18*:295 (1986).

45. Y. L. Lee and B. Crist, *J. Appl. Phys., 60*:2683 (1986).

46. D. Fauteux, Ph.D. Thesis, INRS-Énergie, Université du Québec, P.O. Box 1020, Varennes, Québec, Canada (1986).

47. P. R. Soerensen, RISOE Nat. Lab., RISOE-R-498, DK, 1984, p. 103.

48. P. R. Soerensen and T. Jacobsen, *Electrochem. Acta, 27*:1671 (1982).

49. P. R. Soerensen and T. Jacobsen, *Solid State Ionics, 9*:1142 (1983).

50. D. Fauteux, J. Gauthier, A. Bélanger, and M. Gauthier, The Electrochemical Society Spring Meeting, Montreal, 1982, Abstract 711. Available from the authors.

51. S. Bhattacharja, S. W. Smoot, and D. H. Whitmore, *Solid State Ionics*, *9*:1155 (1986).

52. M. Mali, J. Roos, D. Brinkmann, W. Gorecki, R. Andreani, C. Berthier, and M. B. Armand, Proceedings XXIII Congress AMPERE, Zurich, 1986, p. 240. Available from the authors.

53. J. R. MacCallum, A. S. Tomlin, and C. A. Vincent, *Eur. Polym. J.*, *10*:787 (1986).

54. M. B. Armand, M. Duclot, and P. Rigaud, *Solid State Ionics*, *3*:429 (1981).

55. C. A. C. Sequeria, J. M. North, and A. Hooper, *Solid State Ionics*, *13*:175 (1984).

56. C. A. C. Sequeria and A. Hooper, *Solid State Ionics*, *9*:1131 (1983).

57. F. Bonino, B. Scrosati, and A. Selvaggi, *Solid State Ionics*, *18*:1050 (1986).

58. D. Fauteux, *Solid State Ionics*, *17*:133 (1985).

59. D. Fauteux and M. Gauthier, Proc. of the Symp. on Lithium Batteries (San Diego, 1986), Vol. 87-1, A. N. Dey ed., The Electrochemical Soc., Pennington, New Jersey 08534, p. 514.

60. M. B. Armand, D. Muller, M. Duval, and P. E. Harvey, U.S. Pat. No. 4,578,326.

61. J. R. Giles, J. Knight, C. Booth, R. H. Mobbs, and J. R. Owen, PCT No. WO 86/01643.

62. I. M. Ward, J. E. McIntyre, D. J. Bannister, and P. J. Hall, PCT No. WO 85/02713.

63. K. Nagaoka, H. Naruse, and I. Shinohara, *J. Polym. Sci.*, *22*: 659 (1984).

64. Du Wi Xia and J. Smid, *J. Polym. Sci.*, *22:617 (1984)*.

65. M. Leveque, Doctoral Thesis, Engineering Dept., University of Grenoble, 1986.

66. A. Killis, J. F. LeNest, and H. Cheradame, *Makromol. Chem. Rap. Commun.*, *1*:595 (1980).

67. D. André, J. F. LeNest and H. Cheradame, U.S. Pat. 4,357,401. Nov. 2, 1982.

68. A. Killis, J. F. LeNest, H. Cheradame, and A. Gandini, *Makromol. Chem.*, *183*:2835 (1982).

69. M. Leveque, J. F. LeNest, A. Gandini, and H. Cheradame, *Makromol. Chem.*, *Rapid Commun.*, *4*:497 (1983).

70. A. Killis, J. F. LeNest, A. Gandini, H. Cheradame, and J. P. Cohen-Addad, *Solid State Ionics*, *14*:231 (1984).

71. M. Watanabe, J. Ikeda, and I. Shinohara, *Poly. J.*, *15*:175 (1983).

72. M. Watanabe, J. Ikeda, and I. Shinohara, *Polym. J.*, *15*:65 (1983).

73. M. Watanabe, M. Togo, K. Sanui, N. Ogata, T. Kobayashi, and Z. Ohtaki, *Macromolecules*, *17*:2908 (1984).

74. M. Watanabe, K. Sanui, and N. Ogata, *Macromolecules*, *19*:815 (1986).

75. A. Killis, J. F. LeNest, A. Gandini, and H. Cheradame, *J. Polym. Sci.*, *19*:1073 (1981).

76. A. Killis, J. F. LeNest, A. Gandini, H. Cheradame, and J. P. Cohen-Addad, *Polym. Bull.*, *6*:351 (1982).

77. J. F. LeNest, Doctoral Thesis, University of Grenoble, 1985.

78. M. Armand and D. Muller, U.S. Pat., 4,579,793, April 1986.

79. M. Armand, D. Muller, and J. M. Chabango, U.S. Patent, 4,620,944, Nov. 1986.

80. P. M. Blonsky and D. F. Shriver, *J. Am. Chem. Soc.*, *106*: 6854 (1984).

81. H. R. Allcock, P. E. Austin, T. X. Neenan, J. T. Sisko, P. M. Blonsky, and D. F. Shriver, *Macromolecules*, *19*:1508 (1986).

82. P. M. Blonsky, Ph.D. Thesis, Northwestern University, 1986. Available from University Microfilms International, 300 N. Zeeb Road, Ann Arbor, Michigan 48106.

83. D. J. Bannister, M. Doyle, and D. R. Macfarlane, *J. Polym. Sci.*, *23*:465 (1985).

84. D. Fish, I. Khan, and J. Smid, *Polym. Prepr.*, *27*:325 (1986).

85. S. E. Bell and D. J. Bannister, *J. Polym. Sci.*, *24*:165 (1986).

86. I. Kelly, J. R. Owen, and B. C. H. Steele, *J. Electroanal. Chem.*, *168*:467 (1984).

87. I. C. Hardy and D. F. Shriver, *J. Am. Chem. Soc.*, *107*:3823 (1985).

88. R. Spindler and D. F. Shriver, *Macromolecules*, *19*:347 (1986).

89. G. E. Pacey, *Lithium Crown Ether Complexes*, R. O. Bach, ed., John Wiley, New York, 1985, p. 35.

90. N. K. Dalley, *Synthetic Multidentate Macrocyclic Compounds*, R. M. Isatt, ed., Academic Press, New York, 1978, p. 207.

91. S. Clancy, D. F. Shriver, and L. A. Ochrymoycz, *Macromolecules*, *19*:606 (1986).

92. M. Watanabe, M. Rikukawa, K. Sanui, K. Ogata, H. Kato, T. Kobayashi, and Z. Ohtaki, *Macromolecules*, *17*:2902 (1984).

93. J. R. MacCallum, M. J. Smith, and C. A. Vincent, *Solid State Ionics*, *11*:307 (1984).

94. J. S. Tonge and D. F. Shriver, *J. Electrochem. Soc.*, *134*:270 (1987).

95. M. B. Armand, *Current State of PEO-Based Electrolytes*, Elsevier Appl. Science, C. Vincent and J. MacCallum, eds., London, 1987, p. 1.

96. M. B. Armand, and F. Elkadiri, Proc. of the Symp. on Lithium Batteries (San Diego, 1986). A. N. Dey, ed., Vol. 87-1, The Electrochemical Soc., Pennington, New Jersey, p. 502.

97. G. Vassort, M. Gauthier, P. E. H. Harvey, F. Brochu, and M. Armand, Proc. of the Symp. on Lithium Batteries (Honolulu 1987), Vol. 88, A. N. Dey, ed., The Electrochemical Soc., Pennington, New Jersey 08534. To be published in 1988.

98. N. Kobayashi, T. Hamada, H. Ohno, and E. Tsuchida, *Polym. J.*, *18*:661 (1986).

99. M. Watanabe, S. Nagano, K. Sanui, and N. Ogata, *Polym. J.*, *18*:809 (1986).

100. A. Hooper and J. M. North, *Solid State Ionics, 9*:1161 (1983).

101. M. Gauthier, D. Fauteux, G. Vassort, A. Bélanger, M. Duval, P. Ricoux, J. M. Chabagno, D. Muller, P. Rigaud, M. B. Armand, and D. Deroo, Second Int. Meeting on Li Batteries, Paris, France, April 1984. See also: M. Gauthier et al., *J. Electrochem. Soc.*, *132*:1333 (1985).

102. P. Ricoux, M. Gauthier, and M. B. Armand, Electric Vehicle Int. Symp. AVERE, Versailles, France, June 1984, comm. EII-2. Available from the authors.

103. A. Hooper, J. S. Lundsgaard, and J. R. Owen, Comm. Eur. Communities (Rep.) No. EUR 8660, 1984, p. 53–67. See also R. M. Dell, A. Hooper, J. Jensen, T. L. Markin and F. Rasmussen, p. 68–80.

104. B. B. Owens, Report to EPRI, R.P. 370-30, Aug. 18, 1986.

105. M. Gauthier, A. Bélanger, D. Fauteux, P. E. Harvey, B. Kapfer, M. Duval, C. Robitaille, and G. Vassort, Proc. of the Symp. on Lithium Batteries (San Diego 1986), Vol. 87-1, A. N. Dey, ed., The Electrochemical Soc., Pennington, New Jersey 08534, p. 525.

106. A. Bélanger, M. Gauthier, B. Kapfer, M. Duval, G. Vassort, P. E. Harvey, and E. Saheb, 70th Canadian Chemical Conference, Québec, June 1987. Available from the authors.

107. M. Gauthier, Int. Symp. on Polymer Electrolytes, St. Andrews, UK, June 1987. Available from the authors.

108. G. Vassort, P. Ricoux, M. Gauthier, P. E. Harvey, and F. Brochu, Extended Abstracts of the 3rd Int. Meeting on Li Batteries, Kyoto, June 1986, p. 238.

109. D. J. Barker, Power Sources Symposium, Cherry Hill, N.J., June 1986. Available from the authors.

110. M. Gauthier, A. Bélanger, D. Fauteux, M. Duval, B. Kapfer, M. Robitaille, R. Bellemare, and Y. Giguere, Extended Abstracts of the 3rd Int. Meeting on Li Batteries, Kyoto, June 1986.

111. B. Kapfer, M. Gauthier, G. Vassort, P. E. Harvey, and F. Brochu, Int. Symp. on Power Sources, Brighton, UK, Sept. 1986. Available from the authors.

112. A. Bélanger, M. Gauthier, M. Duval, B. Kapfer, C. Robitaille, M. Robitaille, Y. Giguere, and R. Bellemare, Int. Symp. on Polymer Electrolytes, St. Andrews, U.K., June 1987. Available from the authors.

113. F. Tudron, "Rechargeable Lithium Inorganic and Organic Batteries," Third Int. Seminar on Li Battery Tech. and Applic., Florida, March 1987. Available from Lithium Battery Conference Coordinator, Shawmco Inc., 14 Kawes Road, Sudbury, Mass. 01776.

114. P. Baudry, M. Armand, M. Gauthier, and J. Massounave, Proc. 6th Int. Conf. on Solid State Ionics Symp., Garmisch, Germany, Sept. 1987. To be published in Solid State Ionics in 1988.

115. C. A. C. Sequeria and A. Hooper, The Electrochemical Society Spring Meeting, San Francisco, June 1983, Extended Abstract 493.

116. T. A. Skotheim and O. Inganas, *J. Electrochem. Soc.*, *132*: 2116 (1985).

117. O. Inganas, T. A. Skotheim, and S. W. Feldberg, *Solid State Ionics, 18 & 19* (1986).

118. O. Inganas, I. Lundstrom, and T. A. Skotheim, *Handbook of Conducting Polymers*, T. A. Skotheim, ed., Marcel Dekker, New York, 1986, Ch. 16.

119. T. A. Skotheim, M. I. Florit, A. Melo, and W. E. O'Grady, *Phys. Rev. B30*:4846 (1984).

120. B. Marsan, Ph.D. Thesis, INRS-Energie, Université du Québec, P.O. Box 1020, Varennes, Québec, Canada, 19 6.

121. C. M. Lampert, *Solar Energy Materials, 11*:1 (1984).

122. R. Defendini, M. Armand, W. Gorecki, and C. Berthier, Electrochemical Society Fall Meeting, San Diego, Oct. 1986, Abstract 597.

123. Y. Hirai and C. Tapi, *Appl. Phys. Lett., 43*:704 (1983).

124. B. Kapfer, Doctoral Thesis, University of Grenoble, 1982.

125. B. Kapfer, M. Gauthier, A. Bélanger, and M. Robitaille, IREQ Report No. 8RT3358C.

126. J. Owens, Workshop on Conducting Polymers—Special Applications, Sintra, Portugal, July 1987. Available from the authors.

127. L. C. Hardy and D. F. Shriver, *J. Am. Chem. Soc., 108*: 2887 (1986).

128. D. Muller, J. F. LeNest, H. Cheradame, J. M. Chabagno, and M. Leveque, Eur. Patent, 021 3985, Nov. 1987.

3
Electrode Modification with Polymeric Reagents

Héctor D. Abruña / Cornell University, Ithaca, New York

I. INTRODUCTION

The deliberate control of reactivity at the electrode/solution interface represents one of the primary objectives of electrochemical research. Such control, if possible, would profoundly affect the
areas of electrocatalysis, corrosion, analysis, electrochromics, and
many others. The one variable that has always been available to
electrochemists is the applied potential. By controlling it, one can
dictate the energetics of the interface. However, this affords only
modest variations and generally poor selectivity, especially when
compared to the range of reactivities available through chemical
means. Thus, investigators have pursued the deliberate modification of the electrode surface as a means of dictating and controlling
its properties. Through the deliberate design of the modification
process, it is hoped that the physicochemical properties of the modi-
fier will be transferred to the electrode surface, thus giving it the
range of reactivity and selectivity available in homogeneous chemical
systems. In addition, and perhaps more importantly, the preparation
of structured and molecularly designed interfaces could give rise to
new reactivities and novel devices.

The field of chemically modified electrodes (in its various forms)
thus seeks to dictate and control the properties of the interface
through its deliberate modification.

The intent of this chapter is to present this area of electrochemical research with emphasis on redox polymers and extended
structures. However, since the modification of electrodes with monolayers serves as a good point of departure, these will be considered
as well, although briefly.

I will assume that the audience will not necessarily be well-
versed in electrochemistry, and so a short introductory discussion on
some electrochemical concepts will be presented.

It is hoped that the concepts presented here will be valuable
to electrochemists and non-electrochemists alike who wish to join this
most interesting area of research.

The outline to be followed will be:

1. Brief historical perspective
2. Basic electrochemical concepts and techniques as applied to
    chemically modified electrodes
3. Survey of procedures employed in the modification of electrode
    surfaces
4. Selected examples and applications

The intent is not to give a comprehensive review of the literature,
but rather to present an introduction with emphasis on redox polymers and to point to some of the new frontiers that await this interesting and promising area of research.

There have been a number of reviews [1–7] that have emphasized various aspects, including polymer-modified electrodes [8], in addition to the comprehensive and authoritative review by Murray [1], which covers the literature to about 1982.

II. HISTORICAL PERSPECTIVE

I should mention at the outset that the field of redox polymers is not of recent vintage. In fact Cassidy's book [9], published in 1965, clearly expresses the existence of and interest in such materials. Numerous redox polymers are described in this monograph, including materials containing quinones, ferrocenes, vinylpyridinium, phenothiazine, and others. This early work focused on the synthesis and characterization of redox polymers, but they were not employed as materials with which to modify the surface of an electrode.

The origins of the field of chemically modified electrodes and redox polymers can be traced to the areas of adsorption, chemisorption, and electrodeposition (electroplating). The literature of these areas is truly vast, and will not be addressed in any detail. In the early days of these fields, the approaches taken were essentially empirical and, as a result, there was no coherent framework that could bridge experiment with theoretical prediction. A significant step in trying to unravel some of these processes, especially those related to adsorption at electrode surfaces, was provided by the early work of Fred Anson in the 1960s and early 1970s. From his work emerged a framework for the rational and systematic understanding of the chemical and structural principles underlying adsorption [10].

The first example of the deliberate modification of an electrode surface by exploiting concepts of chemisorption was by Lane and Hubbard [11,12] who chemisorbed a variety of olefin-bearing species onto platinum electrodes, thus exploiting the well-known propensity of olefins to chemisorb onto platinum. Both electroactive and nonelectroactive species were incorporated, including various hydroquinones with an olefin-bearing sidechain, pentahaloplatinates (Br$^-$, Cl$^-$) with coordinated allylamine, 3-allylsalicylic acid, and others. There were a number of interesting observations. One of the more significant ones was that the ability of adsorbed 3-allylsalicylic acid to coordinate iron from solution depended on potential, with binding taking place at potentials negative of 0.0 volts, but not at potentials positive of +0.2 volts. This was interpreted in terms of the potential of zero charge (E_{pzc}) of platinum being about −0.1 volts in the medium involved. (Note: The potential of zero charge represents a condition in which the electrode and the solution potentials are the same, so that there is no net charge at the electrode.) This was an especially notable observation since it demonstrated the ability to control the chemical reactivity of a surface-confined group via the applied potential.

These early observations generated much enthusiasm and stimulated further investigations. In 1975, L. Miller and coworkers reported the preparation of a "chiral electrode" [13] via the immobilization of an asymmetric reagent (S(-)phenylalanine methyl ester). When used in the electrolysis of 4-acetylpyridine, optically active products (albeit with small enantiomeric excesses) could be obtained. This was a clear indication that, by suitable modification of the surface, reactions could be made to proceed via a predetermined pathway.

Also in 1975, Murray and coworkers, in their now classic paper [14], demonstrated for the first time how a designed synthetic sequence could be employed for the deliberate modification of an electrode surface. In this work, Murray and coworkers made use of the reactivity of organosilanes, commonly employed in the preparation of bonded phases for chromatographic applications, and applied them to the modification of tin dioxide electrodes, since these were known to have reactive surface hydroxy groups. They immobilized a number of species and were able to follow the course of the synthetic steps by employing ESCA (electron spectroscopy for chemical analysis). A significant advance was made when they found that superficially oxidized metal electrodes (e.g., Pt and Au) had similar surface hydroxy groups and were also amenable to surface modification [15]. The use of silane-based chemistry was then extended to the incorporation of numerous redox couples [16]. Most of these procedures gave rise to electrodes modified with amounts in the monolayer (ca. 1×10^{-10} moles/cm^2) range. This early work by Murray clearly demonstrated the feasibility of modifying the surface of an electrode in a rational and deliberate fashion.

A versatile extension of this silane chemistry by the Wrighton group at MIT was to incorporate a redox center on a hydrolytically unstable silane reagent so that upon reaction with surface hydroxy groups, the surface would be modified by the specific reagent. The initial materials contained a ferrocene unit [17,18], but Wrighton's group has prepared a broad repertoire of molecules that incorporate a variety of redox groups. These materials were notable because they gave rise to surfaces modified with large amounts of material (coverages as high as 10^{-6} moles/cm^2) and, in addition, these coatings were stable to repeated cycling between oxidation states. The original intent was to use these reagents to stabilize small band gap semiconductors (e.g., Si) employed in photoelectrochemical cells against photoanodic corrosion. In fact, these reagents proved to be quite effective for this purpose [19].

At about this time, Miller and coworkers [20] reported the preparation of redox polymers containing aromatic nitro groups, quinones, and other redox active components, and on their use as materials with which to modify the surface of electrodes.

Oyama and Anson [21] made a significant observation when they reported that electrodes coted with protonated films of poly-4-vinyl pyridine (PVP) could incorporate large amounts of electroactive anions such as $[Fe(CN)_6]^{-4}$. This techniques of electrostatic binding opened a new venue for research in chemically modified electrodes since, by virtue of its simplicity, it allowed for numerous systems to be prepared and studied. This approach was extended to the use of polyanionic materials, most notably Nafion [22] (a perfluoro sulfonate membrane produced by DuPont) and polystyrene sulfonate [23].

The work of Miller and Anson marked the transition from electrodes modified with monolayers (or a few monolayers) to what would be called polymer-modified electrodes.

In 1981, Murray, Meyer, Abruña and coworkers [24] reported that vinylpyridine and vinylbipyridine complexes of ruthenium and iron could be readily electropolymerized to give rise to adherent and electrochemically active polymeric films of the parent monomer. The reaction proved to be quite general and a large number of electropolymerizable metal complexes were prepared and many applications explored. Of these, the so called bilayer-electrode [25,26] represented an especially interesting device in that it could exhibit rectifying behavior without the use of any semiconductor material.

The utility of structured electrochemical interfaces was further demonstrated by the groups of Wrighton [26] and Murray [27], who were able to prepare electronic devices through the use of spatially well-defined layers of polymers.

These areas continue to grow at a very rapid rate, fueled by the belief that virtually any redox material can be incorporated. Thus, new strategies and new applications continue to appear.

Concurrent with these developments were investigations on electronically conducting polymers such as polyacetylene, polypyrrole, polythiophene, polyaniline, and many others. This is an area that has also experienced an explosive growth in terms of new materials and applications [28]. However, this area is largely beyond the scope of this review and as such will not be covered to any extent.

III. BASIC ELECTROCHEMICAL CONCEPTS

A. Cyclic Voltammetric Responses for Freely Diffusing Species and Surface-Immobilized Monolayers

Although the emphasis of this chapter is on the preparation and use of redox polymers, is worthwhile to consider briefly the electrochemical responses of redox centers immobilized on an electrode surface (i.e., a modified electrode) vs. those obtained with freely diffusing species in solution. The discussion will center around cyclic voltammetry, since it is the technique that has most often been employed in these studies. Readers who are well-versed in electrochemistry may feel justified in going on to the next section.

In a cyclic voltammetric experiment [29], one has an unstirred solution containing supporting electrolyte and a redox species (in solution or on the surface). The experiment consists of applying a triangular wave form (rates typically varying from a few millivolts to a few volts per second) to an electrode with the concurrent measurement of current. A plot of current vs. potential is termed a voltammogram.

Consider a solution of a species Ox which can be reduced to Red, with a characteristic potential $E^{o'}$ for the process. We begin by applying a potential E_{init} (1 in Fig. 1) that is sufficiently removed from $E^{o'}$, that essentially no current flows. We will initiate a scan in the negative direction. As the applied potential approaches $E^{o'}$ the current begins to increase (2) and continues to do so (3) until a peak (4) is reached. Upon continued scanning, the current decays in a characteristic "tailing" fashion (5). At point (6) (E_{SW}) the direction of the sweep is changed and the reverse sweep will be

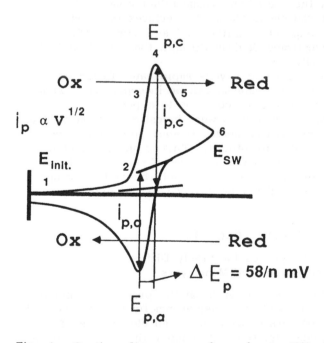

Fig. 1   Cyclic voltammogram for a freely diffusing species.

analogous to the forward sweep (as long as E_{sw} is $>35/n$ mV, n = number of electrons transferred, past the peak). Qualitatively, we can describe the shape of this voltammogram by considering that early on the applied potential is so far removed from $E^{o'}$ that no current flows. In the vicinity of $E^{o'}$ we begin to reduce (Ox to Red) the material that is closest to the electrode surface and at point (4) we essentially exhaust it so that the surface concentration of Ox is zero. Any further material to be electrolyzed needs to diffuse to the electrode surface from the solution and this gives rise to the tailing decrease in the current (often referred to as diffusional tails). The reverse sweep can be understood in analogous terms except that the process now is the reoxidation of the reduced material that was generated in the forward sweep. In a cyclic voltammogram the parameters of importance are the peak potential ($E_{p,c}$ and $E_{p,a}$) (c = cathodic, a = anodic) and current ($I_{p,c}$ and $I_{p,a}$) values as well as the difference in the peak potential values ΔE_p. For a chemically reversible system, (that is, when Ox and Red are both chemically stable species), the ratio of peak currents will be unity at all sweep rates. For an electrochemically reversible system, (that is, one that is always at Nernstian equilibrium), the difference in peak potential values will be 58/n mV (at 25°C). Deviations from these conditions indicate complications, either kinetic or chemical. (See Bard and Faulkner [29] for an authoritative treatment of these.) The average of the peak potentials is taken as the formal potential ($E^{o'}$) for the redox couple. The peak current (in amp) is given by the Randles-Sevcik equation:

$$i_p = (2.69 \times 10^5) \, n^{3/2} \, A \, D^{1/2} \, C^b \, v^{1/2} \qquad (1)$$

where n is the number of electrons transferred, A is the electrode area (cm^2), D is the diffusion coefficient (cm^2/sec), C^b is the bulk concentration (moles/cm^3) and v is the sweep rate (v/sec). Thus i_p is proportional to $v^{1/2}$, C^b, and $n^{3/2}$.

The response obtained at an electrode coated with a monolayer of an electroactive material is shown in Fig. 2. It is immediately apparent that the response (often referred to as a surface wave) is quite different from that previously described. We can go through an analogous analysis. Again, at potentials far removed from $E^{o'}$ (1) no current flows. The current increases as the potential is scanned negatively (2,3) and again a current peak (4) is observed. However, in this case the current decay (5) is just as rapid as the increase was, giving rise to a symmetrical response. The reverse scan is a mirror image of the forward scan. This behavior is due to the fact that in this case there is no diffusion of electroactive material to the electrode since it is all already at the electrode. The voltammogram shown is also for a chemically and electrochemically

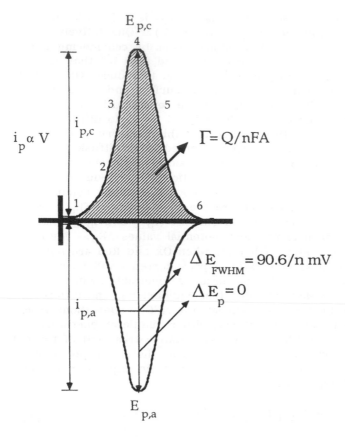

Fig. 2   Cyclic voltammogram for a surface-immobilized redox couple.

reversible system. As for the diffusion case, a chemically reversible system will exhibit a ratio of peak currents of 1. However, for an electrochemically reversible system the ΔE_p value will be zero with larger values (which increase with sweep rate), indicating some sluggishness in the electron transfer event. According to the theory for surface immobilized redox centers [11,30], if surface concentration and activity are equated, the current/potential behavior in a cyclic voltammetric experiment for a reversible process is given by:

$$i = \frac{n^2 F^2 \Gamma v}{RT} \frac{\chi}{(1 + \chi)^2} \tag{2}$$

or:

$$i = - \frac{4i_p \chi}{(1 + \chi)^2} \tag{3}$$

where $\chi = \exp [(nF/RT)(E - E^O)]$, Γ = total amount of reactant initially present on the electrode surface, and v = potential sweep rate. The peak current is given by:

$$i_p = \frac{n^2 F^2 \Gamma v}{4RT} \tag{4}$$

From this it follows that the peak current will be linearly dependent on the rate of potential sweep. This is important to note because it means that the signal-to-noise ratio will not degrade with increasing sweep rate, as is the case for solution species. This is due to the fact that the double layer charging current also grows linearly with sweep rate.

There are, in addition, other properties of surface waves that have no equivalent in the cyclic voltammetric response for diffusing species. The first one has to do with the fact that the oxidation or reduction of the surface layer represents a coulometric experiment. Thus, by integrating the area under the voltammetric wave (i.e., obtaining the charge Q) one can determine the surface coverage (Γ):

$$\Gamma = \frac{Q}{nFA} \tag{5}$$

(Note: for most of the materials to be considered in this review, a monolayer will represent approximately 1×10^{-10} moles/cm^2.)

Finally, for the case of non-interacting neighbors within a film, the full width at half-maximum (FWHM) for a surface wave should be 90.6/n mV. Values larger than this reflect repulsive interactions, whereas smaller values indicate attractive or stabilizing interactions [31–34]. The limit of extremely sharp peaks usually indicates a phase transformation.

In general, many immobilized redox couples exhibit symmetrical wave shapes as well as the linear dependence of peak current with rate of potential sweep. The ΔE_p values are typically small (<30 mV) but seldom zero and generally do not increase with increasing sweep rate, indicating that there are no kinetic limitations. However, the true origin of this effect is not at present clear.

Also, although many of the reported voltammograms are symmetrical, values of FWHM are typically larger than expected. One interpretation [31,34,35] for this effect is based on the assumption that in a layer of immobilized redox centers, there might be (and more than likely are) subtle variations in structure, solvation, environment, etc. that give rise to a range of closely spaced $E^{0'}$ values (rather than a singular one) and that this is responsible, at least in part, for the observed broadening.

Alternatively, this has been interpreted in terms of surface activity coefficients that are dependent on surface coverage. This approach, proposed by Brown and Anson [33] and later modified by Murray and coworkers [34], introduces the concept of an interaction parameter r_i, whose properties (positive values indicate attractive interactions with narrowed peaks, whereas negative values indicate repulisve interactions with broadened waves) dictate the width of the wave. This leads to a modified version of the wave equation:

$$i = \frac{-4i_p \exp(P)}{[1 + \exp(P)]^2 - (r_{Ox} + r_{Red})\exp(P)} \qquad (6)$$

where $P = \frac{nF}{RT}(E - E^o) + (r_{Ox} + r_{Red})\Gamma_{Ox} - r_{Red}\Gamma_{Tot}$ and where it is generally assumed that $r_{Ox} = r_{Red} = 2r$.

B. Polymer Films

Polymer films represent an environment that is significantly different from monolayer films. First of all, surface coverages can range from a few monolayers up to thousands of monolayers, and the local concentrations of redox species can be in the range of 0.1–0.5 $\underline{M}$, which is much greater than values typically encountered in solution.

Because there are numerous aspects to be considered, reactions at these interfaces are more complex than those at electrodes covered with monolayers. These aspects include the mechanisms for electron transfer and transport through the film, especially to sites that are physically remote from the electrode surface. Also of importance is the movement of polymer chains to accommodate the movement of species through the film, including charge-compensating ions as well as solvent molecules (Fig. 3). The exact voltammetric waveshape is dictated by the interplay of all these factors. As a result, a wide range of voltammetric waveshapes have been reported [36–52].

Kaufman et al. [40] proposed a model for the transport of charge through polymer films in which charge propagation takes place via electron self-exchange reactions between oxidized and reduced neighbors within the film. However, although this may represent the mechanism for the process, as we mentioned previously, the overall rate (and therefore the wave shape) will be dictated by the combination of the effects mentioned above. We can consider the diffusional

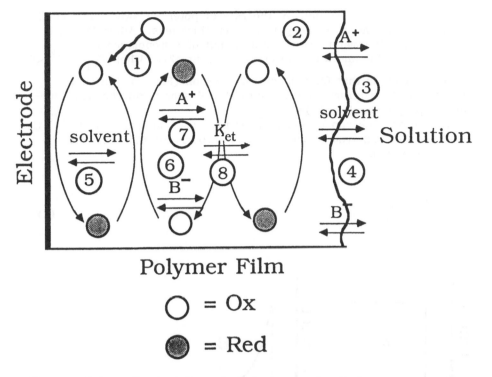

Fig. 3   Schematic depiction of various aspects of charge propagation through polymer films containing electroactive groups.   The processes depicted include: polymer motions (1); transfer of cations (2), solvent (3), and anions (4) across the polymer solution interface; transport of solvent (5), anions (6), and cations (7) through the polymer film; and electron transfer between redox partners (8).

and surface-wave behavior (the latter is also often referred to as "thin film behavior") as representing the two limiting cases of voltammetric shapes.   Whether one, the other, or a combination of the two is dominant will depend on the specific details of the systems under study and the timescale of the experiment involved.

An approach that has been quite successful in the treatment of this problem is based on the fact that transport rates through these films appear to obey the laws of diffusion [53–82].   Thus, the measurement of diffusion coefficients—variously termed D_{ct} (charge transport diffusion coefficient), D_{app} (apparent diffusion coefficient), or D_{eff} (effective diffusion coefficient)—has provided some very useful insights into these processes.   This coefficient is usually expressed in terms of the parameter $D_{ct}\tau/d^2$, where τ is the time scale of the experiment (dictated by the rate of potential sweep in a cyclic voltammogram) and d is the polymer film thickness. When D_{ct} τ/d^2 is $\gg 1$, the voltammetric waveshape is that of a monolayer in Nernstian equilibrium with the electrode (Fig. 2). When D_{ct} $\tau/d^2 \ll$ 1 the response is that for a species in solution (Fig. 1). In this case the voltammogram loses its symmetrical shape and the peak current is proportional to $(v)^{1/2}$.

In addition, one can describe these processes in terms of $D_{ct}^{1/2}$ C, which is obtained from a potential step experiment (Fig. 4) and the Cottrell equation:

$$i = \frac{nFAD_{ct}^{1/2}C}{\sqrt{\pi}\,\sqrt{t}} \tag{7}$$

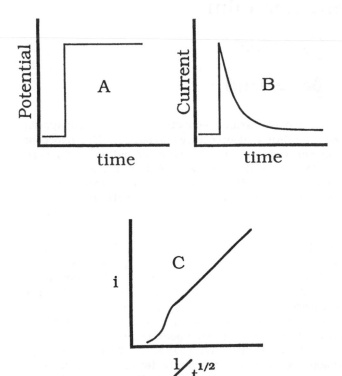

Fig. 4   Potential step experiment (A), current/time decay (B), and Cottrell Plot (C) for an electrode covered with a redox polymer.

From the value of the slope at short experimental timescales (Fig. 4), one can obtain the value of $D_{ct}^{1/2} C$. It is important to use the slope at short times because in the Cottrell equation semi-infinite linear diffusion is assumed; at long experimental timescales the diffusion profile can reach the outer boundary of the polymer film, so this assumption is no longer valid. For systems with large values of D_{ct}, at slow experimental times, or for sufficiently thin films, the voltammetric response will again be that of a monolayer film with symmetrical waveshapes and peak currents linearly proportional to the rate of potential sweep. For the opposite case of small D_{ct}, thick films, or fast experimental times, the response will revert back to that exhibited by a solution species with the characteristic diffusional tails.

In many cases one can observe the transition from a surface-wave response to a diffusional wave response by simply changing the rate of potential sweep. The literature of polymer-modified electrodes is replete with examples of this.

In some instances the anodic and cathodic waveshapes can be significantly different. This is particularly prevalent in systems where one of the oxidation states of the polymer is electrically neutral. In addition, solvent effects can be dramatic in such instances.

Perhaps the most thorough study of waveshapes and their changes was by Peerce and Bard [43], who studied and simulated the electrochemical response for electrodes modified with polyvinylferrocene (see Fig. 5).

In the limit of strong interaction between near neighbors, extremely sharp waveshapes, indicative of a phase transition type of behavior, can be observed. One of the more illustrative examples of this was the study by Bard and Henning [83,84] on tetrathiafulvalene (TTF) in Nafion.

There have also been some attempts at trying to factor out the various contributions to measured values of D_{ct}. For example, Majda and Faulkner [85] employed a luminescence quenching technique to measure the apparent diffusion coefficient of $[Ru(bpy)_3]^{2+}$ in polystyrene sulfonate in the absence of counterion motion. The measured value was larger than that obtained by potential step methods (which necessitates counterion movement), indicating that the latter process is rate-determining in the transport of charge.

Buttry and Anson [68] measured the apparent diffusion coefficients for $[Co(bpy)_3]^{3+/2+}$ and $[Co(bpy)_3]^{2+/1+}$ electrostatically bound to Nafion. Due to the very large difference in the electron self-exchange rates for these two processes (with the former being larger by a factor of more than 10^7), they were able to separate the contribution of electron self-exchange from the apparent diffusion coefficient. This is dramatically demonstrated in Fig. 6, which shows

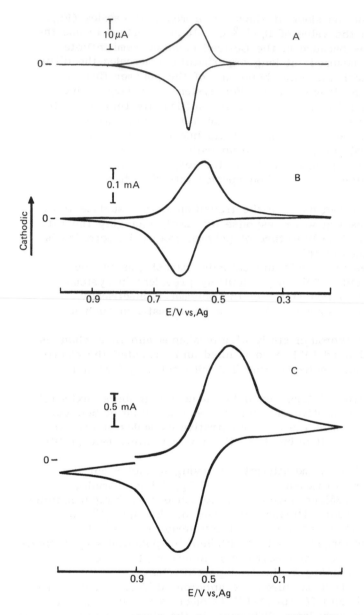

Fig. 5   Cyclic voltammograms in 0.1M TBAP/MeCN for a platinum electrode covered with a film of polyvinyl ferrocene. Scan rates: A = .01 V/sec, B = 0.2 V/sec, C = 10 V/sec. (From Ref. 43, used with permission.)

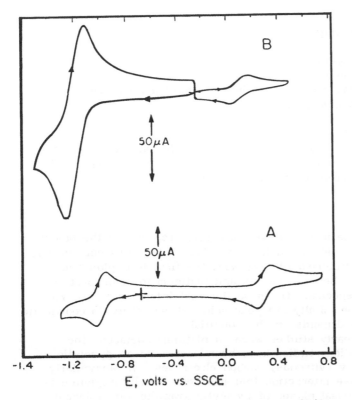

Fig. 6   Cyclic voltammograms for $[Co(bpy)_3]^{3+/2+}$ and $[Co(bpy)_3]^{2+/1+}$ in solution (A) and incorporated into Nafion (B). (From Ref. 68, used with permission of the American Chemical Society.)

a cyclic voltammogram for both processes in solution and in Nafion. The very large difference in peak currents in the latter case is a manifestation of the contribution of the electron self-exchange process to the apparent diffusion coefficient of $[Co(bpy)_3]^{2+/1+}$, but not to that of $[Co(bpy)_3]^{3+/2+}$. These observations were in good accord with the theoretical treatments of Dahms and Ruff [86].

IV.   METHODS FOR MODIFYING THE SURFACE OF AN ELECTRODE WITH REDOX-ACTIVE COMPONENTS

A variety of methods can and have been employed in incorporating redox components on electrode surfaces including:

1. Adsorption
2. Covalent bonding
3. Polymer films
   a. Redox polymers
      1. Organic redox couples
      2. Redox centers containing transition metal ions
   b. Ion-exchange polymers
4. Others

The first two areas will be considered briefly since they provide a good point of departure. However, most of the discussion will center on the preparation of redox polymer films.

A. Adsorption

As mentioned in the introduction, adsorption represents the oldest and simplest technique for anchoring redox centers to electrode surfaces. Lane and Hubbard [11,12] were the first to exploit the chemisorptive bond between olefins and platinum to incorporate a number of redox species. Of these the most interesting case was that of the adsorbed 3-allylsalicyclic acid, whose ability to coordinate iron from solution depends on the potential.

Although these early studies were on platinum surfaces, the majority of adsorption studies have been performed on pyrolytic graphite and glassy (vitreous) carbon electrodes. A particularly strong and effective interaction that has been heavily exploited is that between the basal planes of pyrolytic graphite and molecules having extended pi electron systems. Examples of these are found in the early work of Anson and coworkers, who adsorbed phenanthrene quinone, o-dianisidine, a $Ru(NH_3)_5Py$ (Py = pyridine) moiety bound to phenanthrene, and others [33,87].

The vast majority of adsorption studies carried out on carbon surfaces have been directed at electrocatalytic applications, especially the reduction of molecular oxygen. A wide range of substituted porphyrins and phthalocyanines have been investigated. Of these, the most thorough and successful series of investigations have been by Anson and Collman [88–96]. Of particular note is the now classic example of O_2 reduction employing face-to-face porphyrins [88,89,93].

In a related area, there has been a recent surge in the use of Langmuir-Blodgett films for the modification of electrode surfaces. A variety of redox couples have been incorporated this way, including ferrocenes [97a–c], bipyridine complexes of ruthenium and osmium, [97d,e], viologens [97b,c], and others [97f].

B. Covalent Bonding

Silane-Based Modification

Although adsorption represents the simplest way to modify the surface of an electrode, it suffers from the fact that it is an equilibrium process. Thus, inevitably, the material will desorb from the electrode surface, giving rise to electrodes with limited lifetimes. In addition, adsorption rarely gives rise to coverages in excess of a monolayer, thus limiting the amount of material that can be incorporated.

These limitations prompted the investigation of alternative means of modifying electrode surfaces. It was Murray's group who first articulated a general strategy for the covalent attachment of redox centers onto electrode surfaces. The approach was based on the use of silane reagents that had been employed by chromatographers in the derivatization of silica surfaces for the preparation of bonded phases. Such reagents could be reacted with the surface hydroxy groups found on metal oxide electrodes or superficially oxidized metals such as Pt and Au to give rise to covalently attached species via siloxane bonds. This generated much enthusiasm and a wide range of redox active centers were immobilized via this approach [14,98-103]. One of the most versatile reagents for this application was the so-called ethylenediamine silane (3(2-aminoethylamino)-propyltrimethoxy-silane), whose primary amino end-group could be employed to attach a wide range of species containing either a carboxylic acid or an acid chloride group. In these early days, a great deal of emphasis was placed on establishing the surface chemistry and thus, ESCA [98-110] was extensively employed to follow the course of such reactions.

Another approach based on silane chemistry was to attach a silane that contained a coordinating group and subsequently react this "surface ligand" with a metal complex containing a weakly bound and easily displaced ligand. Such ligand substitution reactions have been carried out with immobilized pyridine nitrile and thio ligands [111].

Murray's review [1] contains an extensive listing of materials immobilized via silane chemistry.

In most of these examples, the electrochemical responses follow those anticipated for immobilized redox reagents. In addition, the formal potentials for the immobilized species follow closely those of the dissolved analogs. This indicates that immobilization does not significantly alter electronic differences between the oxidized and reduced forms, which dictate the formal potential. This also has the important consequence of bringing in an element of predictability, which can be (and has been) exploited in electrocatalytic applications.

Modification of Carbon Electrodes

Given the immense literature of organic synthesis, the modification of carbon surfaces represents a most logical step. Most electrode modification studies on carbon have focused on pyrolytic graphite and glassy carbon, with carbon paste electrodes receiving some attention. The modification of carbon electrodes has been largely based on the manipulation of the surface groups present on the basal planes. These are generally oxygenated functionalities such as alcohols (phenols), ketones (quinones), and carboxylic acids.

A number of approaches have been employed in the modification of carbon surfaces by exploiting the known chemistry of the groups present on the edge plane. One of the most successful has been the generation of acid chloride groups by reaction of the electrode with thionyl chloride. Subsequent treatment with an amino-substituted species gives rise to covalent bonding via amide bond formation. The first example of this was Miller's chiral electrode [13]. In addition, a wide range of metalloporphyrins based on tetra(4-aminophenyl) porphrin have been immobilized in this fashion. These include the free base as well as the Fe, Co, Cu, Zn, Ni, and Mg metalloderivatives [112–115]. Various ferrocenes [116] and ruthenium complexes (NH_2-Py-Ru(NH_3)$_5$ [117], -PyRu(EDTA) [118]) have also been incorporated. In general, however, coverages obtained were low.

Kuwana [119–122] introduced an approach based on the reactivity of cyanuric chloride with surface hydroxy groups. With this approach he was able to incorporate various redox couples, including hydroxymethyl ferrocene [119,120], a viologen (methyl, aminopropyl viologen) [121], as well as a phenylene-diamine [122]. Fox employed this procedure for the incorporation of quinones [123] and Yacynych for immobilizing glucose oxidase [124].

Carbon paste electrodes present a rather different medium for modification [125–133]. In this case, the material of interest is often mixed with the pasting material, so that the use of the term immobilization is not always applicable. The biggest advantage of this approach is the easy regeneration of the electrode. A number of applications of this approach have appeared, including electrocatalysis [129a,130] and electroanalysis [129b,131–133]. It is my belief that the use of carbon-paste-modified electrodes will in the future find widespread analytical use.

C. Modification with Polymer Films

Although the early work on chemically modified electrodes centered on materials at about monolayer coverages, most of the recent effort has been centered on polymers. This is partly due to the fact that the signals obtained at electrodes modified with polymer films are greatly amplified over those at monolayer coverages, thus making their measurement much easier. For catalytic applications polymer films are expected to exhibit higher stability. Also, there were theoretical predictions [160] that pointed to the advantages of polymer films (over monolayers) in electrocatalytic applications. In addition, their use opened the way for the preparation of deliberately structured interfaces. Furthermore, many of the methods employed in the modification step can be very simple and include dip coating, spin coating, electropolymerization, and plasma polymerization.

Many approaches to the preparation of electroactive polymer films have appeared. These can be subdivided into two main categories depending on how the redox-active species is immobilized. If the redox center is part of the polymer backbone it is termed a redox polymer. The second group is termed ion-exchange polymers, and in this case the redox active component is a counterion to a polyionic (anionic or cationic) film. One can also differentiate preformed polymers that are subsequently applied to an electrode surface from polymeric films that are generated in situ.

As mentioned previously, due to the typically larger coverages involved, the electrochemical response of polymer films can be greatly complicated by the interplay of charge transfer and transport, polymer film motions, solvent swelling and other processes. As a result, the voltammetric responses obtained are much more difficult to interpret, and, in fact, a great deal of effort has been focused on the unravelling of these processes [36-85].

The preparation of electrodes modified with redox polymers can be achieved in a number of ways. One can employ a preformed polymer containing the redox center, or the redox center can be coupled to a previously prepared polymer (e.g., by ligand substitution). They can also be generated in situ by the electroinitiated polymerization of redox active monomers.

Ion exchange polymers serve as charge-compensating ions for the incorporation of redox species. These materials can be prepared by placing an electrode modified with the polymeric ion exchange material in a solution of the redox ion, from which the polymer extracts the redox ions. Alternatively, the ion exchange polymer can be deposited with the charge-compensating redox ion already incorporated.

Redox Polymers

Discussion of the preparation of redox polymers will be divided into systems that are based on organic redox couples and those containing transition metal complexes. They will be further subdivided into preformed polymers and polymers generated in situ.

Preformed Redox Polymers Containing Organic Redox Couples.
The preparation of polymeric materials bearing organic redox-active
components is attractive since the entire synthetic repertoire can be
applied. However, because of stability considerations, care must be
exercised in the choice of redox systems. The most commonly employed ones have been quinones, viologens, and aromatic nitro groups
because of their well-behaved electrochemical responses and synthetic
versatility. As mentioned previously, Cassidy's monograph has an
extensive discussion on the preparation of such materials, although
none of them were employed in the modification of electrode surfaces.

One of the earliest reports of electrodes modified with preformed
polymer films bearing an organic redox-active component was by
Miller and coworkers. They prepared polymers containing aromatic
nitro groups [20a,b,c], quinones [20d,e,f,135], dopamine [136], and
other redox-active components. All of these were bulk polymers,
and so they were purified and characterized prior to use. They
were deposited onto platinum or graphite electrodes either by adsorption or droplet evaporation of dilute solutions. The adhesion and
stability of these films were greatly enhanced by a brief thermal
treatment that apparently increased the degree of crosslinking.

Miller and Degrand [20d] prepared and studied a particularly well-
characterized family of polymers derived from the condensation of
2-anthraquinonecarbonylchlroide and poly(ethylenimine). The degree
of loading of the quinone was varied from 47% to 75%. These polymers
were deposited on carbon and mercury electrodes, and their redox
behavior was studied. A wide range of responses were obtained,
depending on the coverage, solvent properties, and degree of crosslinking. For example, the pH dependence of the response of modified carbon electrodes is presented in Fig. 7. In the case of
mercury electrodes [20f], the polymers were adsorbed onto a hanging
mercury drop electrode (HMDE) and the redox response obtained as
the drop was expanded. Dramatic differences in the response were
obtained (see Fig. 8). In all of these studies, Miller and coworkers
emphasized (correctly) the importance of polymer chain motions on
the observed electrochemical responses.

They have employed these polymer modified electrodes for a number
of applications, including the electrocatalytic reduction of aromatic
halides (employing a poly nitrostyrene film) (20a,c] and the oxidation of NADH (employing a polymer with bound dopamine) [136]. An
especially elegant application was the use of a polymer which upon
reduction cleaved to release a neurotransmitter in what resembles a
primitive synaptic response (via infra) [137,140].

Kaufman, Schroeder and coworkers have studied a number of polymers derived from poly(vinyl benzylchlroide) that contain TTF
(tetrathiafulvalene), ferrocene, or pyrazoline derivatives as redox-
active components [38,40,141-145]. (See Fig. 9.) These materials

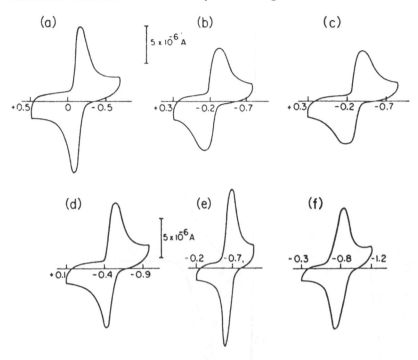

Fig. 7   Cyclic voltammograms at 0.1 V/sec and at different pH for
a graphite electrode modified with an anthraquinone polymer.   pH
values are: (a) 0.7, (b) 2.0, (c) 4.0, (d) 7.0, (e) 10.00, and
(f) 13.   (From Ref. 20d, used with permission.)

are attractive because of their well-behaved redox responses as well
as potential structural effects that might arise (e.g., TTF/tetra-
cyanoquinodimethane (TCNQ)).   These performed polymers were
characterized prior to use and then deposited onto platinum electrodes—
from THF solutions—by spin coating.   The electrochemical responses were obtained in acetonitrile, which is a nonsolvent for
these polymers.   A number of studies focused on the use of the TTF
derivative and on the effects of polymer morphology, degree of
crosslinking, and electrolyte type on the redox behavior.   The
studies found that the electrochemical behavior depends strongly on
the electrolyte type and that significant changes in polymer film
volume and morphology accompany electrochemical oxidation.

   In collaboration with Chambers [40], these researchers also studied
the ion and electron transport dynamics through these films.   From
this study, they proposed a model for the transport of charge through
polymer films whereby charge propagation takes place via electron

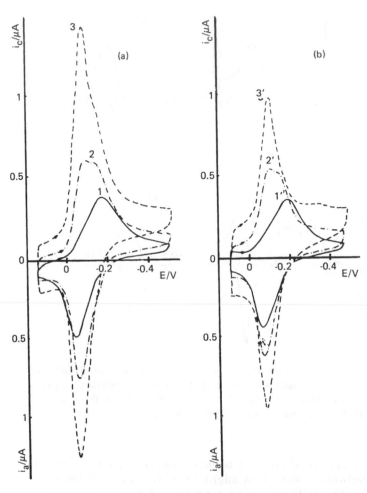

Fig. 8 Influence of drop area on the voltammograms of an anthraquinone polymer adsorbed on a mercury electrode. Voltammetry in
(a) aqueous and (b) alcoholic solution; mercury drop area (in mm^2):
1 and 1': 1.0; 2 and 2': 2.1; 3 and 3': 4. (From Ref. 20f,
used with permission.)

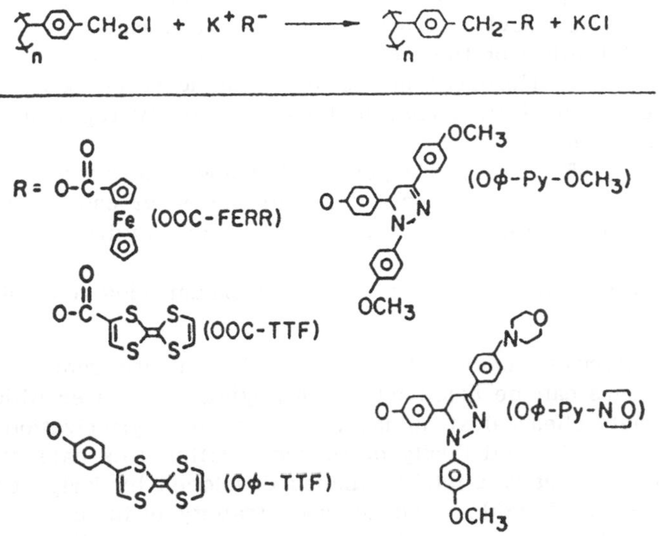

Fig. 9   General synthetic method for the preparation of functionalized polystyrenes bearing electroactive pendant groups.   (From Ref. 38, used with permission.)

self-exchange reactions between oxidized and reduced neighbors within the film.

Chambers and coworkers [51–53,146–152] have performed extensive studies on TCNQ–containing polymers with particular emphasis on the variation of the chemical composition and molecular structure of the polymer.   They have studied the effects of solvent, counterion type and concentration.   In addition, they have performed potential-dependent spectroelectrochemical as well as ESR experiments on the modified electrodes.   From these experiments, they were able to detect the presence of a mixed valence species.   In addition, they observed an equilibrium mixture of radical ion sites $(TCNQ^{\overset{\bullet}{-}})_x$ and a dimeric form $(TCNQ^{-2})_x$ in fully reduced films.   This is an example of the novel reactivity patterns that can be elicted due to the very high local concentration of redox centers.

Funt and Hoang [153] prepared and characterized a family of polymers of poly(vinyl p-benzoquinone) and copolymers with styrene. In their electrochemical studies they noted that whereas electrodes modified with the homopolymer exhibited asymmetric waveshapes, those modified with copolymers showed more symmetrical ones.   This was ascribed to repulsive interactions between adjacent quinone sites. In addition, very pronounced solvent effects were noted and these were attributed to the ability of a solvent to swell the polymer film.

Degrand and Laviron [154] prepared a family of copolymers derived from p-phenylazobenzoyl chloride and poly(ethylenimine) in
which the degree of loading of the electroactive azobenzene was
varied from 5% to 95%. Differences in the voltammetric responses
for various degrees of loading, similar to those previously reported
by Miller, were observed.

Abruña and Bard [155] employed a polyxylyl-viologen to modify
the surface of platinum and p-Si electrodes. In the latter case, colloidal platinum was incorporated and, upon illumination, hydrogen
was evolved.

Rosenblum and Lewis [156] have also employed similar viologen polymers on Pt and n-Si.

Organic Redox Polymers Generated In Situ. The in situ generation of redox polymers can be achieved by employing a monomer which,
under the appropriate chemical stimulus, will undergo polymerization.
Perhaps the best-characterized family of reagents within this category
is the group of hydrolytically unstable silanes developed by Wrighton
and coworkers [17-19,157-180]. The general strategy is to incorporate hydrolytically unstable silyl groups onto a redox-active
molecule. Upon hydrolysis, a stable polymer based on siloxone
linkages is formed. A wide variety of reagents have been prepared
by this approach (see Fig. 10), with those containing ferrocene and
viologen groups having received the most attention. In addition,
anthraquinone and tetraalkylbenzenediamine derivatives have been
prepared. In this section we will focus on those systems containing
organic redox couples.

The viologen derivatives (D and E in Fig. 10) have been studied
most extensively because of their versatility [163,164,167-172,176-
180]. P-type silicon electrodes have been modified with these
viologens, with the subsequent incorporation of colloidal platinum or
palladium. Upon irradiation of the semiconductor with light of
energy exceeding the band gap, the photoassisted evolution of hydrogen (photovoltages of about 0.5 V) was achieved [163,164,167,168,
170]. Electrodes modified with the viologen polymers and with colloidal
palladium were also active in the catalytic reduction of bicarbonate
to formate [171,172,176].

Platinum electrodes modified with viologens were effective in the
mediated reduction of cytochrome C [165]. In addition, viologen
films deposited on transparent tin oxide electrodes exhibited a well-
behaved electrochromic response [168,169].

In addition to the above applications, Wrighton and coworkers have
studied the rates of charge transport through the viologen polymer
films [177,180]. In these studies they demonstrated that the +/0
couple had a significantly higher value of D_{ct} than the +2/+1 couple.
Substantial changes (30-40%) in the polymer film thickness were observed upon reduction from the 2+ to the 0 oxidation state (see Fig.
11).

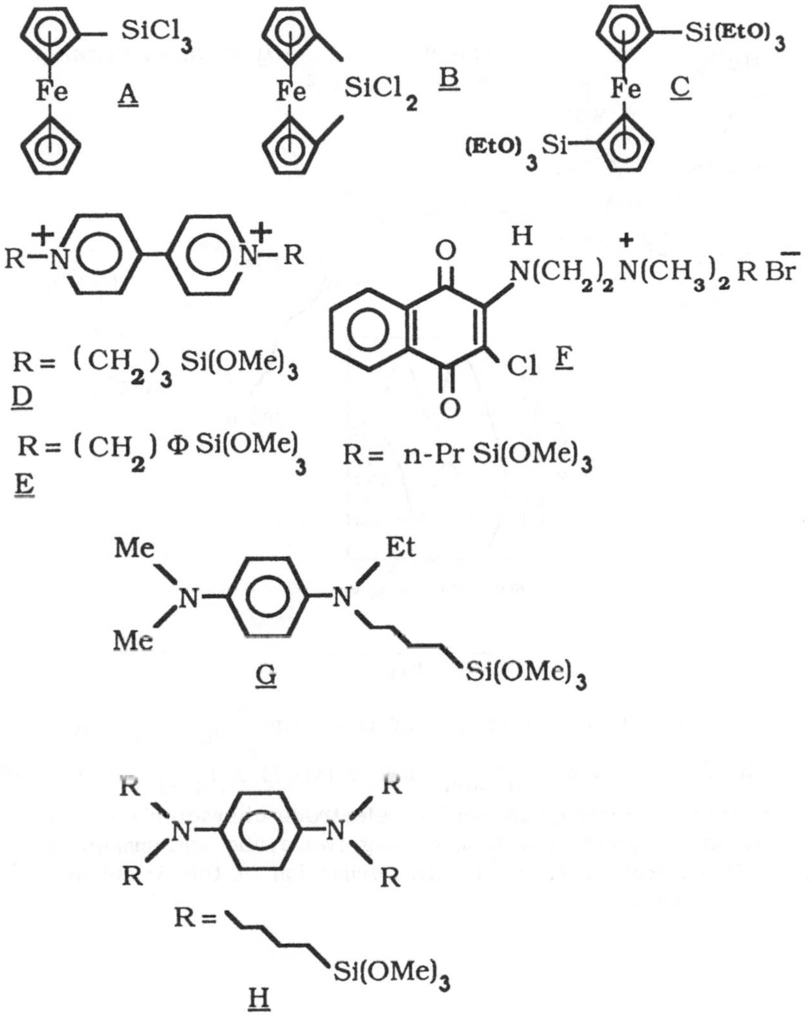

Fig. 10 Hydrolytically unstable silanes employed by Wrighton and coworkers.

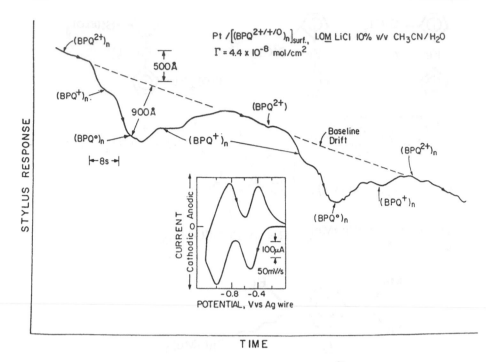

Fig. 11 Change in the thickness of the $[(BPQ^{2+})_n]_{surf}$ polymer upon reduction to $[(BPQ^+)_n]_{surf}$ and to $[(BPQ^0)_n]_{surf}$. The inset shows a cyclic voltammogram for the electrode characterized. The plot of stylus response vs. time during two cyclic voltammograms is shown. (From Ref. 177, used with permission of the American Chemical Society.)

The naphthoquinone derivative (F in Figure 10) was used to modify both platinum and p-WSe2 electrodes [173]. Well-developed voltammograms were obtained at both electrode materials, although in the case of p-WSe2 illumination was required. The derivatized Pt electrode exhibited a response whose pH dependence (56 mV/pH unit) was very close to the predicted value of 59. The modified platinum and p-WSe2 electrodes were active in the electrochemical and photoelectrochemical reduction of oxygen to hydrogen peroxide, respectively. The immobilized benzene diamines (G and H in Fig. 10) [174] exhibited well-behaved electrochemical responses for the first oxidation, but generation of the dication resulted in rapid loss of activity presumably due to hydrolysis. These modified interfaces were active in the mediation of ferro/ferri cytochrome C. These studies are illustrative of the wide range of reactivities that are accessible through reagent immobilization.

Elliott and coworkers [181] prepared derivatives of $[Ru(byp)_3]^{2+}$ with acrylate substituents on the bipyridine ligand which could be subsequently thermally polymerized. Such modified electrodes exhibited outstanding electrochromic behavior with seven reversible color changes [181c]. [Note: although these studies were on transition metal complexes they were included here because the deposition procedures are closely related to those described in this section.]

An alternative approach to in situ generation of redox polymers is electropolymerization. Although a number of redox couples have been immobilized in this fashion, the exact nature of the polymerization process has often not been well-defined. Alberry and coworkers [182] prepared polymeric films of thionine (about 20 monolayers) on platinum and tin oxide electrodes. The modified electrodes exhibited a broad voltammetric response centered around 200 mV. These researchers carried out an extensive study of the photogalvanic properties of such modified electrodes in the presence of a wide variety of redox couples in solution.

Haas and Zumbrunnen [183] have reported the anodic electropolymerization of hydroxyphenazine on glassy carbon electrodes from aqueous media and the ability of modified electrodes to mediate a number of redox processes, such as the reduction of $[Ru(bpy)_3]^{3+}$, Fe^{3+}, quinoxaline, and O_2.

Moutet [184] reported the anodic electropolymerization of vinyl triphenyl amine and its use for the electrocatalytic oxidation of carboxylate anions.

Dubois [185] and coworkers electropolymerized 5-hydroxy-1,4 napthoquinone onto graphite electrodes. In this case the polymerization mechanism was observed to be similar to that exhibited by phenol: Initial oxidation gives rise to a radical and propagation takes place via head-to-tail coupling, followed by deprotonation.

Torstensson and Groton [186] employed a procedure similar to that of Alberry and were able to prepare graphite electrodes modified with sulfonated phenothiazines and to employ them for the catalytic oxidation of NADH.

Deronzier and coworkers [187] prepared a pyrrole-substituted viologen derivative and exploited the well-known ability of pyrrole to undergo oxidative electropolymerization to prepare electrodes modified with a polymeric viologen film. It is interesting to note that whereas polypyrrole films are insulating at potentials negative of about -0.2 V, the modified electrodes exhibited two well-defined reductions, although the waveshapes were significantly different (Fig. 12). In addition the 2+/1+ couple appears to have faster charge transfer kinetics than the +1/0 couple, and this is in contrast to the previously mentioned results of Wrighton and coworkers.

Murray and coworkers [188] and Elliott and coworkers [189] have prepared vinyl viologens which undergo electroreductively initiated

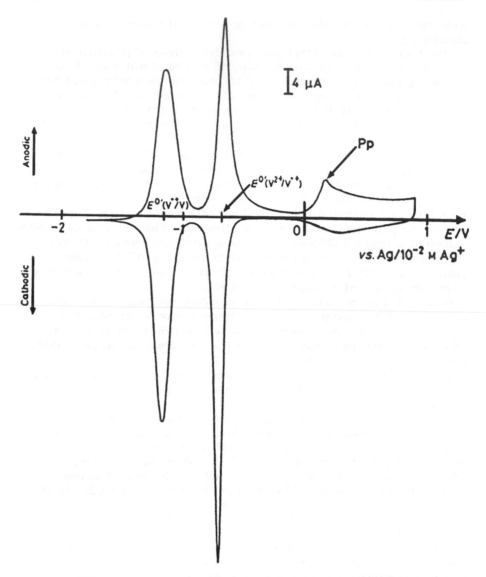

Fig. 12 Cyclic voltammogram for a poly (pyrrole-viologen)-modified electrode in 0.1 M TBAF/MeCN. (From Ref. 187, used with permission.)

polymerization to give rise to electrodes coated with adherent and electrochemically active films of the viologen.

Most of the studies on organic redox couples incorporated by ion exchange have been on viologen derivatives and these will be discussed in the section of ion exchange incorporation of transition metal complexes.

The preparation of polymers containing organic redox reagents continues to be very attractive because of all the synthetic variations that are possible and because the materials may be fully characterized. It also allows for the synthesis of elaborate materials with which engineered microstructures can be prepared on electrodes.

Preformed Redox Polymers Containing Transition Metal Ions. Perhaps the first electrochemical study of a redox polymer containing a transition metal was by Oyama and Anson [190], who employed pyrolytic graphite electrodes coated with polyvinylpyridine or polyacrylonitrile and made use of the coordinative properties of the pyridine and nitrile groups to incorporate ∿Py-Ru(EDTA, ∿CNRu-(EDTA), ∿PyRu(NH$_3$)$_5$, and ∿CNRu(NH$_3$)$_5$ (∿Py and ∿CN refer to polymer-bound pyridine and nitrile groups, respectively). After the initial observation that such ligand exchange reactions could be performed, they carried out a series of experiments in which the coordination chemistry of the immobilized complexes was probed [191]. Particular emphasis was placed on the Ru(EDTA) complex coordinated to a PVP-modified electrode. First of all, the observed electrochemical response was a function of the oxidation state of the ruthenium present in solution from which incorporation took place. When ruthenium was present as [Ru(III)(EDTA)], a well developed wave for ∿PyRu(III)EDTA could be observed at an E$^{o'}$ value of approximately −0.15 V. However, when the solution contained Ru(II)EDTA, waves for both ∿PyRu(EDTA) and ∿Py$_2$Ru(EDTA) could be observed. The fact that two pyridines could coordinate with Ru^{II} whereas only one did with Ru^{III} is in excellent concurrence with the higher degree of π backbonding present in Ru(II). More dramatic was the ability to take an electrode modified with ∿Py$_2$RU(EDTA) and observe the loss (reversible) of a coordinated pyridine upon oxidation to RuIII. Spectroelectrochemical experiments were also performed to follow the dissociation kinetics [192].

Oyama and Anson were also able to prepare electrodes in which two different redox centers are coordinated to the same surface ligand (see Fig. 13) and in which the same complexes are coordinated to two surface ligands.

More recently, Anson's group prepared a copolymer containing pyrrolidone and iron porphyrin groups, which they applied to the reduction of O$_2$ [193]. The advantages of this system were that,

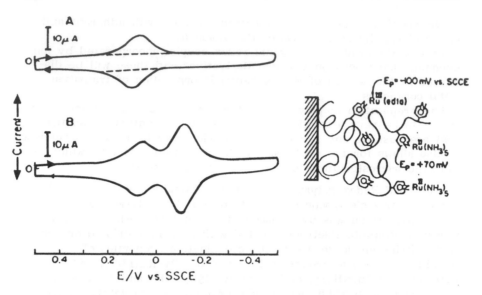

Fig. 13   (A) Steady-state cyclic voltammogram for a PVP-coated
electrode that was soaked for 15 min in 10 mM $Ru(NH_3)_5OH_2^{2+}$,
washed, and transferred to a supporting electrolyte of 0.2 M
CF_3COONa (pH 4.2).   Scan rage: 200 mV/sec.   The dashed line
is the curve obtained before soaking.   (B) Voltammogram resulting
when the electrode from A was soaked for 2.5 min in 0.1 mM
Ru^{III}(EDTA), washed, and transferred to the same supporting
electrolyte.   Scan rate: 200 mV/sec.   (From Ref. 191, used with
permission of the American Chemical Society.)

whereas the polymer was insoluble in aqueous solutions, the hydrophilic character of the pyrrolidone groups gave rise to good swelling
characteristics.   Although electrocatalytic activity was observed, it
was not very long-lived.

Meyer's group at North Carolina has applied its extensive synthetic
expertise to the preparation of a wide range of polymeric materials
containing complexes of Ru, Os, Re, and other transition metals.
The early work focused on complexes coordinated to PVP.   An
especially well-characterized series of polymers was based on the reaction of $[Ru(terpy)(bpy)(H_2O)]^{2+}$ (terpy is 2,2':6'2" terpyridine)
with PVP to give rise to $[Ru(terpy)(bpy)(\sim py)]^{2+}$ in which the total
loading of material was varied systematically [194].   At low loadings,
the spectral features of the polymer were essentially identical to those
of the monomer.   However, at higher loadings, significant changes
were observed, indicating interactions between the chromophores
along the polymer chain.   These researchers were also able to

incorporate $[\sim PyRu(bpy)_2(H_2O)]^{2+}$ [195] and exploit its oxidative electrocatalytic activity. Although the stability of the system was not high, catalytic turnovers were clearly established. In addition they noted that at pH values below the pK of the pyridine groups in PVP, the film's pH remained invariant with changes in the solution pH. This effect was termed pH encapsulation [195]. Recent work by Redepenning and Anson [265] has established this effect due to Donnan potential effects.

An especially clever way of incorporating a redox site was by electrochemically generating an open coordination site on a metal complex. If this is done in a noncoordinating solvent (CH_2Cl_2) at an electrode coated with PVP, the metal complex can bind to a pyridine site. This was elegantly demonstrated by using a variety of Os and Re chloro complexes which upon reduction undergo chloride ion loss. The complex is coordinated by a surface pyridine group which in essence "captures" the metal complex [196].

Meyer and Ellis [197] devised a very general approach for the incorporation of metal complexes into preformed polymers. The approach is based on the chemical reactivity of p-chlorosulfonated polystyrene [197] towards amino, hydroxy, and carboxylate groups to give rise to sulfonamide, sulfonester, and sulfonanhydride linkages, respectively. Thus, by employing metal complexes that bear one of the aforementioned substituents on the periphery of one of the ligands, a large number of metal complexes can be immobilized. Since the sulfonamide linkage is the most stable, the incorporation of amino-bearing species has been emphasized. Redox centers incorporated include a wide range of polypyridine complexes of ruthenium, iron and osmium, aminoferrocene, nickeltetraaza macrocycles and others (Table 1). This approach has been employed in various applications, including bilayer electrodes [197], photoactive assemblies [198], luminescent polymers [192], and others [199].

Vos and Haas have prepared electrodes modified with $[\sim PyRu(bpy)_2Cl]^{1+}$ [200-202] and $[\sim Im-Ru(bpy)_2Cl]^{1+}$ [203] ($\sim$Im represents a polymer-bound imidazole from poly n-vinyl imidazole). These electrodes showed well-behaved electrochemical responses, but more important was their photochemical behavior. Upon irradiation, a photosubstitution took place, with the final product being strongly dependent on the solvent and the presence of other ligands (Fig. 14). The presence of isopotential points (these are the electrochemical equivalent of isosbestic points in spectroscopy) indicates a simple reaction sequence. Such reactivity is unusual in that one would anticipate rapid quenching of the excited state by the electrode. Perhaps most of the redox centers are sufficiently removed from the electrode that the quenching is not effective. These results are in contrast to those of Abruña, Murray, and Meyer with a similar system [103] in which no photoactivity was noticed. However, in the latter case, the coverage was near monolayer so that quenching could have been more effective.

Table 1   Complexes Immobilized by Sulfonamide Formation

Redox Couple	$E^{O'}$ soln, V^a	$E^{O'}$ surf, V^a
$(bpy)_2Ru(5\text{-phenNH}_2)^{2+}$	+1.31	+1.28
$(byp)Ru(5\text{-phenNH}_2)_2^{2+}$	+1.39	+1.37
$Ru(5\text{-phenNH}_2)_3^{2+}$		+1.53
$(bpy)_2Ru_2(4\text{-pyNH}_2)_2^{2+}$	+0.99	+1.14
$(bpy)_2Ru(3\text{-pyNH}_2)_2^{2+}$	+1.18	+1.32
$(bpy)_2Ru(4\text{-pyOH})_2^{2+}$	+1.18	+1.20
$(bpy)_2Ru(3\text{-pyOH})_2^{2+}$	+1.24	+1.25
$(bpy)_2Ru[4,4'\text{-bpy(COOH)}_2]^{2+}$	+1.28	+1.25
$(bpy)_2Ru(4\text{-pyCOOH})Cl^+$	+0.75	+0.74
$(NH_3)_5Ru(4\text{-pyNH}_2)^{2+}$	+0.12	−0.18
$Fe(5\text{-phenNH}_2)_3^{2+}$		+1.23
$Fe(5\text{-phenNH}_2)_2(CN)_2$	+0.40	+0.32
$(\eta^5\text{-C}_5H_5)Fe(\eta^5\text{-C}_5H_4\text{-4-C}_6H_4NH_2)$	+0.31	+0.36
$(5\text{-phenNH}_2)_2Os(dppm)^2$		+1.37
$[Ni(NH_2Ethi)_2Me_2[16]tetraeneN_4)]^{2+}$	+1.16	+1.13
$H_2N\text{-4-C}_6H_4N(CH_3)(DMPD)$	+0.78 +0.15	+0.55
nickel tetrakis(o-aminophenyl)- porphyrin	−1.24 −1.74	−1.32 −1.83

^{a}Volts vs. SSCE in $0.1M$ $(E_T)_4N$ ClO_4/CH_3CN.

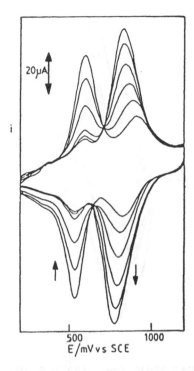

Fig. 14   Cyclic voltammograms for a $[\sim ImRu(bpy)_2Cl]^{1+}$ modified electrode during photolysis in $1M$ $HClO_4$. (From Ref. 203, used with permission.)

In addition to the systems mentioned here, there is an extensive literature on polymer-bound transition metal complexes for catalytic applications [204]. Although many systems have been prepared and investigated, very few have been electrochemically characterized. This is clearly an area that awaits further study and development.

Electropolymerization of Transition Metal Complexes.   Electropolymerization of redox-active monomers represents one of the most versatile ways to modify the surface of an electrode. The process is generally applicable to a broad range of electrode materials, including metals, semiconductors, carbon, and conducting metal oxides. In addition, the coverage can be exquisitely and reproducibly controlled through polymerization conditions.

Perhaps the best-characterized and most extensive family of compounds in this regard are the vinylpyridine, vinylbipyridine (and related compounds) complexes of ruthenium, osmium, and iron first

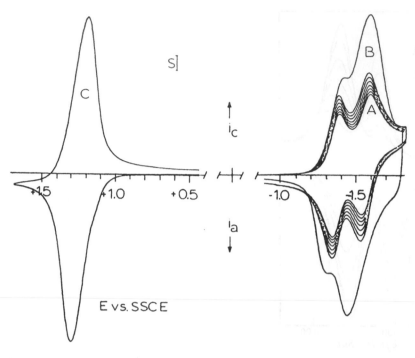

Fig. 15   Cyclic voltammogram in 0.1M TEAP/MeCN for: (A) 0.5 mM
solution of [Ru(bpy)$_2$(v-py)$_2$]$^{2+}$ (first seven consecutive scans)
s = 5μA; (B) electrode modified with a polymeric film of
[Ru(bpy)$_2$(v-py)$_2$]$^{2+}$ s = 25μA; (C) same as (B) but scanning in
the anodic region.   (From Ref. 24, used with permission of the
American Chemical Society.)

reported by Abruña, Murray and Meyer [24].   These materials undergo electroreductively initiated polymerization to give rise to electrodes
coated with extremely adherent, electrochemically active films of the
parent monomer (Fig. 15).   A wide variety of synthetic variations
are possible, and numerous monomers can be copolymerized to give
rise to electrodes with multiple redox states.

Polymerization was believed to be due, in part, to the fact that the
reduction processes are ligand-localized and, in addition, there is a
significant degree of π backbonding, thus localizing a significant
amount of charge on the ligands.   This charge redistribution was
then believed to be responsible for the polymerization, especially in
light of the fact that vinylpyridine undergoes anionic polymerization.
However, extensive studies by Murray and Meyer [205a] indicate that
the dominant mechanism involves radical coupling through the ligands.

Thus, the picture that emerges from these studies is that the polymers are not metallated polyvinylpyridines, but rather ligand-bridged extended structures.

Most recently Guaar and Anson [205b] elegantly corroborated the importance of this pathway by examining the behavior of $[Ru(P)_2(v\text{-}bpy)]^{2+}$ complexes p=phenanthroline derivative). They found that, although polymerization was initiated at the vinyl group, there was an extensive degree of coupling through the 4 and 7 positions of the phenanthroline ligands. In fact, they were able to isolate an ESR-active species that was identified as a ligand-based free radical.

Numerous complexes have been synthesized and electropolymerized following this approach, and these are presented in Table 2. As a rule, all of these materials give rise to exceptionally stable deposits with well-behaved responses and uniform coverage. Virtually pinhole-free films can be obtained for layers as thin as 20 Å [206].

Polymeric films of these complexes have been employed by Murray and coworkers to probe a wide range of processes, including transport phenomena [207], electrocatalysis [208,212–215], redox conduction [27,213] photoactive interfaces [216], and others [217]. In addition, a number of interesting applications have also emerged, including bilayer electrodes [24,25,218], and electronic-type devices [214].

Calvert and coworkers also prepared numerous electropolymerizable complexes derived from pyridineacetylene [219].

Most recently, we were able to prepare and characterize a large series of vinylterpyridine complexes of ruthenium, iron, and cobalt [220]. These materials undergo the predicted electropolymerization to give rise to well-behaved films of the corresponding monomer. In the case of cobalt, novel chemical pathways (especially ligand substitution) not exhibited by the corresponding solution species were found.

Pickup and Osteryoung have studied electropolymerized films of $[Ru(bpy)_2(v\text{-}py)_2]^{2+}$ in room-temperature molten salts [221].

Ellis and Meyer [222] have found that aminophenanthroline complexes of a wide range of metal ions will undergo electrooxidatively initiated polymerization to give electrochemically well-behaved films. As with the previous approach, this method offers great latitude in terms of the synthetic variability. The polymerization is probably analogous to that of aniline. This approach was extended by Spiro and coworkers to a variety of aminoporphyrins [223,224].

Deronzier and coworkers [225] and Parker [226] prepared a series of complexes of pyrrole-substituted pyridines and bipyridines. By taking advantage of the oxidative electropolymerization ability of the pyrrole residue, they were able to deposit a number of metal complexes.

There have also been reports of other electropolymerization procedures, including that of an acylnickel derivative [227], cobalt tetraazaannulenes [228], and Schiff base complexes [229].

Table 2   Complexes Immobilized by Electropolymerization of Vinyl-Containing Ligands

Complex	Soln[a]			Polymer films[a]		
	E°'ox	E°'red(1)	E°'red(2)	E°'ox	E°'red(1)	E°'red(2)
$[Ru(bpy)_2(vpy)_2]^{2+}$	1.24	-1.35	-1.55	1.22	-1.36	-1.52
$[Ru(bpy)_2(vpy)Cl]^{+}$	0.76	-1.49	-1.74	0.76	-1.48	—
$[Ru(bpy)_2(vpy)NO_2]^{+}$	1.03	-1.45	-1.69	1.03	—	—
$[Ru(bpy)_2(vpy)NO_3]^{+}$	0.96	—	—	0.91	—	—
$[Ru(bpy)_2(vpy)NO]^{3+}$	0.53	—	—	0.47		
$[Ru(vbpy)_3]^{2+}$	1.16	-1.43	-1.54	1.16	-1.43	-1.54
$[Fe(vbpy)_3]^{2+}$	0.93	-1.42	-1.58	0.93	-1.44	-1.57
$[Ru(vbpy)_2Cl_2]$	0.30	—	—	0.36	—	—
$[Ru(vbpy)_2(MeCN)vpy]^{2+}$	1.36	—	—	1.29	—	—
$[Ru(vbpy)_2(MeCN)_2]^{2+}$	1.41	—	—	1.35	—	—
$[Ru(trpy)(bpy)(vpy)]^{2+}$	1.21	-1.26	-1.59	1.20		
$[Ru(trpy)(bpy)(BPE)]^{2+}$	1.21	-1.26	-1.58	1.20		
$[Ru(trpy)(bpy)(4'-Cl-stilb)]^{2+}$	1.21	-1.35	-1.55	1.20		

$[Os(bpy)_2(vpy)]Cl^+$	0.35		0.36	-1.46
$[Os(bpy)_2(BPE)]Cl^+$	0.33		0.33	-1.50
$[Ru(bpy)_2(vpy)_2]^{2+}$	1.22	-1.54	1.25	-1.36
$[Os(bpy)_2(vpy)_2]^{2+}$	0.74	-1.53	0.77	-1.33
$[Ru(bpy)_2(BPE)_2]^{2+}$	1.23	-1.53	1.30	-1.35
$[Ru(bpy)_2(stilb)_2]^{2+}$	1.22	-1.54	1.23	-1.36
$[Ru(bpy)_2(4'\text{-}Cl\text{-}stilb)_2]^{2+}$	1.24	-1.54	1.22	-1.38
$[Ru(bpy)_2(4'\text{-}OMe\text{-}stilb)_2]^{2+}$	1.25	-1.59	1.19	-1.39
$[Ru(bpy)_2(4'CN\text{-}stilb)_2]^{2+}$	1.23	-1.64	1.25	-1.40
$[Ru(bpy)_2(p\text{-}cinn)_2]^{2+}$	1.20	-1.70	1.19	-1.39
$[Os(bpy)_2(p\text{-}cinn)_2]^{2+}$	0.72	-1.65	0.74	-1.36
$[Ru(bpy)_2(m\text{-}cinn)_2]^{2+}$	1.27	-1.63	1.28	-1.38
$[Ru(bpy)_2(p\text{-}CH_2\text{-}cinn)_2]^{2+}$		-1.61	1.26	-1.39
$[Ru(bpy)_2(p\text{-}fum)_2]^{2+}$	1.20	-1.71	1.22	-1.40
$[Ru(bpy)_2(N\text{-}Me\text{-}py)_2]^{4+}$	1.19	-1.65	1.22	-1.39

Table 2   (Continued)

Complex	Soln[a]			Polymer films[a]		
	$E°'ox$	$E°'red(1)$	$E°'red(2)$	$E°'ox$	$E°'red(1)$	$E°'red(2)$
$[Ru(phen)_2(vpy)_2]^{2+}$	1.25	−1.37	−1.51	1.24		
$[Ru(trpy)(stilb)_2Cl]^+$	0.77	−1.38	−1.74[b]	0.78		
$[Ru(i\text{-}Pr\text{-}bpy)_2(p\text{-}cinn)_2]^{2+}$	1.46	−0.96	−1.16			
$[Ru(trpy)(vpy)_3]^{2+}$	1.23	−1.24	−1.76[b]	1.22		
$[Ru(trpy)(BPE)_3]^{2+}$	1.23	−1.26	−1.54[b]	1.30		
$[Ru(trpy)(stilb)_3]^{2+}$	1.20	−1.25	−1.68[b]	1.23		
$[Ru(trpy)(4'\text{-}Cl\text{-}stilb)_3]^{2+}$	1.20	−1.25	−1.60[b]	1.22		
$[Ru(HC(pz)_3(vpy)_3]^{2+}$	1.17	−1.58		1.16		
$^*[Ru(py\text{-}py\text{-}py)_2]^{2+}$	1.25	−1.22	−1.44	1.22	−1.24	−1.47
$^*[Ru(py\text{-}py\text{-}py)(typ)]^{2+}$	1.27	−1.22	−1.47	1.24	−1.23	−1.44
$^*[Ru(py\text{-}py\text{-}py)_2]^{2+}$	1.34	−1.20	−1.46	1.37	−1.18	−1.43
$^*[Co(py\text{-}py\text{-}py)_2]^{2+}$	.27	−.75	−1.60	.20	−.80	−1.64

[a]Values vs. SSCE in 0.1M $(ET)_4N/ClO_4/CH_3CN$.
[b]Irreversible reduction, Ep.c. value.   Abbreviations: vpy = 4-vinyl pyridine; v-bpy = 4-vinyl, 4'-methyl, 2,2' bipyridine, BPE = bis pyridine ethylene; stilb - stilbazole, p-cinn = pyridyl cinnamide; p-fum - pyridiyl fummaride; * denotes location of vinyl group in vinylterpyridine.

Ion Exchange Polymers

Cationic Polymers.   As mentioned previously, this method of
electrode modification is based on the principles of ion exchange for
the incorporation of redox species.   This versatile approach was introduced in 1980 by the now-classic papers of Oyama and Anson [21]
in which they termed the procedure "electrostatic binding."   It essentially involves the incorporation of redox-active counterions to an
electrode modified with a polyionic film.   The early work focused on
the use of protonated polyvinylpyridine [21a] and deprotonated polyacrylic acid [21b].   In the former case, polyanionic redox couples
such as $Fe(CN)_6^{3-/4-}$, $Ir Cl_6^{2-/3-}$ were incorporated up to coverages
of about 10^{-7} moles/cm^2 (Fig. 16).   It was noted early on, however,
that factors other than charge control the degree of incorporation.
For example, exposure of a PVP-modified electrode to a 5mM solution
of $Fe(CN)_6^{3-}$ and $IrCl_6^{2-}$ results in greater binding of the latter
[21a].   In addition, $Fe(CN)_6^{3-}$ exhibits a larger partition coefficient
than the more highly charged $Fe(CN)_6^{-4}$.   Thus, there are clearly
numerous aspects involved in the partitioning process.

In the case of deprotonated polyacrylic acid (a polyanion)
$Ru(NH_3)_6^{3+/2+}$ was incorporated, again with high coverage [21b].

In the examples mentioned above, once the electrodes were modified, they could be placed in a solution containing only supporting
electrolyte and a very presistent electrochemical response could be
obtained.   However, since the partitioning is a reversible process,
the redox ions would slowly leach away.   The response of such modified electrodes could be stabilised by having a very small (micro-
molar) concentration of the redox ion in solution.

This approach to electrode modification was extremely attractive
due to its simplicity and broad scope.   After Anson's early reports,
great emphasis was placed on this mode of immobilization.   This was
especially true for multianionically charged redox ions such as
$Fe(CN)_6^{3-/4-}$, $IrCl_6^{2-/3-}$, $Mo(CN)_8^{3-/4-}$, $Ru(CN)_6^{3-/4-}$ and
$Co(CN)_6^{3-/4-}$, which were incorporated into a variety of polycationic films, including quaternized polyvinylpyridine [23,56,59,60,
230,231,234], polymeric viologens [65,163,166,167,232], and others
[233–235].

One of the problems commonly encountered with these materials is
the small apparent diffusion coefficients (typically smaller than 10^{-10}
cm^2/sec).   This has been ascribed to various effects, including poor
swelling of the polymer film by the solvent and the crosslinking effect
that multiply charged ions have.

Since electrocatalytic applications (which represent one of the biggest attractions of chemically modified electrode research) require
rapid rates of charge propagation, investigators have focused on
the development of polymers with open structures in order to accelerate charge transport.   Perhaps the most successful of these has

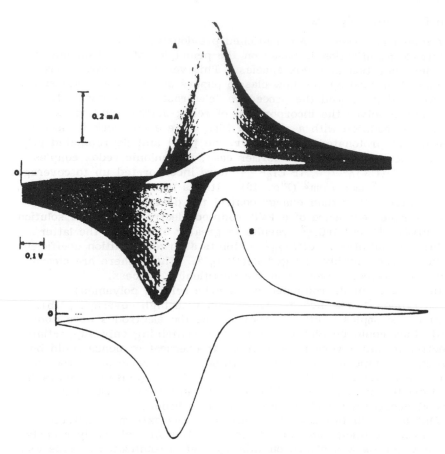

Fig. 16   Cyclic voltammograms for a graphite electrode modified
with PVP and in contact with a 5mM solution of Fe(CN)$_6^{-3}$. (A)
consecutive scans depicting incorporation of complex. (B) voltammogram of the modified electrode in supporting electrolyte 0.2M
TFA, pH = 2.9. (From Ref. 21a, used with permission.)

been poly-l-lysine [66]. This material exhibits exceptional swelling in aqueous media (by as much as a factor of 100) so as to give rise to a polymer with a very open structure and large pools of solvent, allowing for both the electrostatic binding of redox anions and rapid rates of exchange. In fact, some of the largest reported values of apparent diffusion coefficients have been in poly-l-lysine. Anson and coworkers have carried out extensive studies on the transport properties of redox couples incorporated into this material [67,236].

Most recently, Anson presented a thorough study of the electrochemical response of $Fe(CN)_6^{4-/3-}$ incorporated into protonated random and block copolymers containing varying ratios of styrene and p-(diethylaminomethyl) styrene [237]. With ratios of styrene to p-(diethylaminomethyl) styrene groups up to 2:1, the electrochemical responses observed for $Fe(CN)_6^{3-/4-}$ electrostatically bound to coatings prepared from random and block copolymers of the same composition are dramatically different. For block copolymers, large quantities of material can be incorporated, the electrochemical response is very well defined, and the kinetics of charge transport appear to be significantly rapid. The random copolymers, on the other hand, could not incorporate nearly as many redox ions and in addition exhibited very low rates of charge propagation. This, again, points to the critical importance of the polymer structure.

In addition to the materials mentioned, there have been numerous other polycationic polymers employed. In general, the redox systems studied have been the multianionic species such as $Fe(CN)_6^{-4}$, $IrCl_6^{3-}$, $Mo(CN)_8^{4-}$, and related materials [235,238].

Anionic Polymers. One of the earliest anionically charged polymers to be employed was polystyrene sulfonate. Numerous cationic redox species such as $[Ru(bpy)_3]^{2+}$, $[Co(bpy)_3]^{3+}$, and related materials have been studied [85,239]. In a similar vein, Meyer and coworkers have recently employed partially hydrolyzed p-chlorosulfonated polystyrene to incorporate a number of redox species, most of them involving polypyridine complexes of ruthenium and osmium [240,241]. They have employed such films in a number of electrocatalytic applications, including water [240] and chloride oxidation [241].

Perhaps the most widely used anionic polymeric material for modifying electrode surfaces has been the perfluoro sulfonate material Nafion [242]. (See Fig. 17 for structure.) By virtue of the sulfonate side chain, Nafion is a cation exchanger, however its properties are significantly different from those of conventional sulfonate resins (based on divinyl benzene/styrene sulfonate). In conventional sulfonate resins, the ion selectivity properties are dictated by the interrelated effects of ion exchange capacity, degree of crosslinking, and water sorption. Nafion, however, has ostensibly no

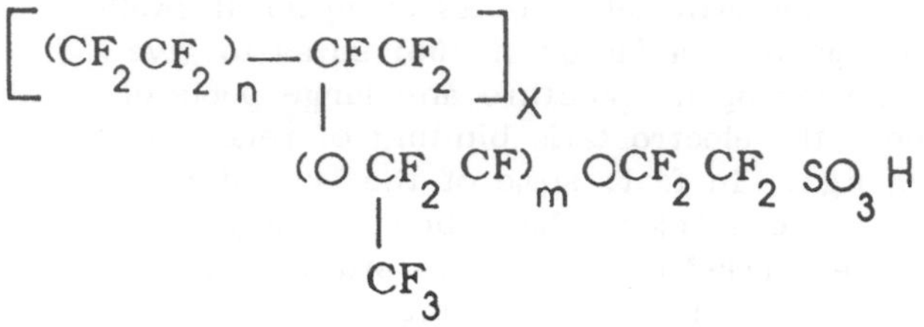

$$\left[(CF_2CF_2)_n \underset{\underset{\underset{\underset{CF_3}{|}}{CF_3}}{\overset{|}{(O\,CF_2\,CF)_m\,OCF_2\,CF_2\,SO_3\,H}}}{\overset{|}{-}CF\,CF_2} \right]_X$$

Fig. 17   Structure of Nafion.

crosslinking, the density of exchange sites is significantly lower than that of typical sulfonate resins, and the fluorocarbon backbone provides a highly hydrophobic environment. Its noncrosslinked nature makes its solvent swelling characteristics very dependent on the nature of the counterion present (and its solvation properties) and pretreatment. In addition, there is mounting evidence that Nafion exists in segregated phases or domains with hydrophobic (fluorocarbon) and hydrophilic (with clusters of sulfonate sites) domains connected by an interfacial region [243]. Nafion also exhibits exceptional affinity for hydrophobic cations. Martin and coworkers [244] have performed extensive studies of selectivity coefficients, and they report values as large as 10^6 (relative to sodium).

Although Nafion is available in a variety of forms, it is the soluble form that has found the most widespread use for electrode modification. Martin and coworkers have reported on a procedure for preparing soluble Nafion from the solid membrane material [245].

Nafion has been extensively employed in numerous studies, especially by the groups of Anson [63,68,246–254], Bard [22,70,71, 83,84,255–260], and Martin [244,245,261–264]. A variety of redox cations have been incorporated into Nafion, including $[Ru(bpy)_3]^{2+}$, $[Fe(bpy)_3]^{2+}$, $[Os(bpy)_3]^{2+}$, $[Ru(NH_3)_6]^{3+}$, methylviologen, $[Co(bpy)_3]^{2+}$, and $[Co(terpy)_2]^{2+}$. Many of the studies have been geared to understanding the details of the charge transport and propagation mechanisms. Anson has also employed these systems for the electrocatalysis of O_2 reduction [247–249]. Bard has focused on the study of charge transport mechanisms in Nafion [83,84]. He has also performed electrochemical/ESR studies of redox species in Nafion [258,259], in addition to using them for catalytic applications [256,257]. Martin and coworkers have carried out extensive studies geared to the understanding of the ion selectivity properties of Nafion films as well as the microenvironments mentioned previously.

One of the problems with Nafion is that materials retained in the hydrophobic domains exhibit sluggish transport rates. This can have deleterious effects in an electrocatalytic cycle if the active component resides mainly in the hydrophobic domain. To remedy this situation, Buttry and Anson [248,249] have developed systems in which an active catalyst is employed in conjunction with a material that will remain largely in the hydrophilic phase and thus serve as an electron shuttle between the electrocatalyst and the electrode.

In addition, Redepenning and Anson [265] have recently reported that the effects of pH and electrolyte concentration on the formal potentials of species incorporated into Nafion layers are due to Donnan potential effects.

Lewis and coworkers [266] have examined the kinetics and mechanisms of substitution reactions of $Ru(NH_3)_5OH_2^+$ with various ligands when the complex is incorporated into a Nafion film. Significant differences in behavior were observed when compared to the behavior in solution.

D. Other Methods of Electrode Modification

There are a number of other methods that do not fit into any of the previously mentioned categories, and these are considered here.

The first one has to do with the use of a plasma discharge to initiate polymerization of vinyl substituted derivatives. The two species that have received the most attention are vinyl ferrocene [37,41,54,55,57,267,268] and vinylpyridine [37,269]. These have been generally deposited onto carbon electrodes (pyrolytic graphite or glassy carbon). Typically, the electrodes are exposed to the Ar^+ plasma, which is believed to clean and activate the surface. Introduction of the reagents and further exposure to the plasma gives rise to polymeric deposits. These are typically adherent films. However, it is difficult to control the film thickness and, in addition, there is generally some damage of the film by the plasma.

The second has to do with precipitated films of polynuclear transition metal cyanides of general formula $M_k^A[M^B(CN)_6]_\ell$ [270]. Of these, prussian blue (iron(III) hexacyanoferrate(II)) is the prototypical example. It is a highly insoluble polymeric inorganic material whose intense color has been known for 250 years. Prussian blue (PB) can be both oxidized and reduced to give Everitt's salt and Berlin green, respectively. The deposition of PB on conducting substrates was first described by Neff [271], who also showed that such films could be electrochemically oxidized. The most extensive studies on PB have been by Itaya, Uchida and Neff who in addition have studied ruthenium purple (ferric ruthenocyanide) and osmium purple (ferric osmocyanide).

Bocarsly and coworkers [272-278] introduced a very elegant method for preparing films of nickel hexacyanoferrate by the dissolution of nickel in the presence of ferricyanide ions. They have performed very extensive studies of this system, including electrochemical characterization and transport studies [272,273] and reflectance studies [275]. In addition, they have studied the dependence of growth on crystallographic orientation [278]. There is an interesting counterion (cation) effect in the redox response of these films which has been ascribed to the ease of penetration of the ions into the lattice as charge compensating counterions during redox transformation.

Finally, there have been numerous applications of these films, including display devices, electrocatalysis and others [249–255].

Majda [286–288] and coworkers have prepared porous alumina films of controlled porosity and have coated the inner walls with various polymers with the subsequent incorporation of redox couples. For example, they have employed protonated PVP and incorporated $[Fe(CN)_6]^{-4}$. In addition, they have employed self-assembling monolayers containing a redox-active species (e.g., viologen). Their main interest is the investigation of lateral electron transfer.

To take advantage of their ion exchange properties, clays and zeolites have also been employed as media for incorporating redox couples [289–296].

V. APPLICATIONS

A. Electrocatalysis

Description

As with any catalytic process, electrocatalysis aims at reducing the energy of activation of a reaction (a redox transformation in this case) by providing low-energy pathways between reactants and products [297]. The ability to do this in a rational and predictable manner was and continues to be one of the main objectives of chemically modified electrode research.

In talking about electrocatalysis at modified electrodes we need to consider electrodes modified with monolayer amounts of material or less and those modified with multimolecular layers. In addition, we need to distinguish between simple charge transfer mediation, whereby the redox centers at an electrode surface serve merely as electron shuttles (that is, ideal outer sphere couples) and the case in which there is a more intimate chemical interaction between catalyst and substrate. These are commonly referred to as redox catalysis and chemical catalysis, respectively.

In the case of redox catalysis at an electrode modified with a monolayer of a redox-active component, Saveant [298] demonstrated that essentially no catalytic effect would be expected, compared to the case of a mediator in homogeneous solution. This was ascribed to a loss of dimensionality. It was also pointed out, however, that redox polymers with 100 monolayer equivalents could be very effective.

In the case of chemical catalysis, catalytic activity could be observed even at monolayer coverages. In addition, the commonly mentioned advantages of using small amounts of catalyst and ready separability of the catalyst from the reaction medium are clearly operational.

A second distinction that needs to be made when comparing reactions at electrodes modified with monolayers vs. multilayers is that, in the latter, the rate of charge propagation [299] and substrate permeability [300,301] can play very important roles, and can in fact be rate determining. Saveant and coworkers [301] have developed a model that incorporates all of these kinetic aspects. They have presented a series of concentration-distance profiles for catalyst sites and substrate within a redox polymer film as a function of the various kinetic parameters (Fig. 18).

Due to the well-defined transport processes and steady state response, electrocatalytic reactions at chemically modified electrodes are often studied by rotated disk electrode techniques. For an electrode covered with a monolayer of material the limiting electrocatalytic current is given by

$$\frac{1}{i_{max}} = \frac{1}{nFAk_{ch}\Gamma C_s} + \frac{1}{0.62nFA\nu^{-1/6}D_s^{2/3}\omega^{1/2}C_s} \tag{8}$$

where Γ represents the surface coverage of catalyst sites, C_s is the solution substrate concentration, D_s is the substrate's diffusion coefficient in solution, k_{ch} is the rate constant for the substrate/catalyst reaction, ν is the kinematic viscosity and ω is the rate of rotation in radians/sec [1]. Plots of $1/i_{max}$ vs. $1/\omega^{1/2}$ (so called inverse Levich Plots or Koutecky-Levich Plots) should be linear, and, from the intercept, the value of k_{ch} can be obtained.

Cyclic voltammetry can also be employed for the study of electrocatalysis, although it is more difficult to extract quantitative information. However, it is simple to establish a qualitative picture. For example, consider an electrode modified with a redox reagent Ox/Red and the reduction of a species Subs. at bare and modified electrodes (Fig. 19). The modified electrode will exhibit a reversible response (curve A), whereas the direct reduction of Subs. takes place at a significant over-potential (that is, at a potential far removed from its reversible potential) (curve B). At the modified electrode, the reduction of Subs. takes place at a potential that is lower than the direct reduction and that, in addition, can be displaced from the potential of the Ox/Red couple depending on the catalytic activity (curve C). If the return wave (anodic in this case) is absent and the peak current is proportional to $v^{1/2}$, the catalysis is rapid and controlled by the diffusion of Subs. to the surface. Although this

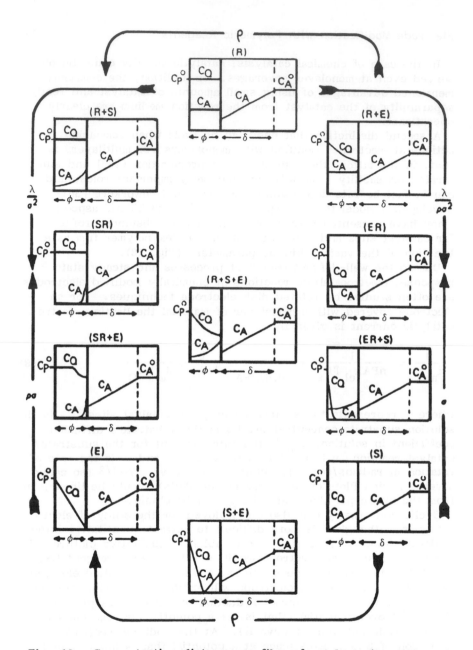

Fig. 18 Concentration distance profiles of catalyst sites (Q) and substrate (A) in a polymer film of thickness φ and Levich layer of thickness δ as a function of various rates.

1. E = electron transport through film
2. S = substrate diffusion
3. R = catalytic reaction

(From Ref. 301, used with permission.)

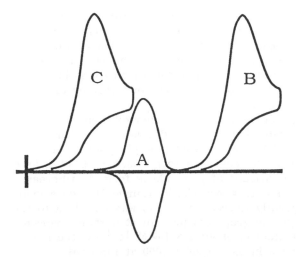

Fig. 19   Depiction of cyclic voltammetric responses for electrocatalysis at a chemically modified electrode. (A) voltammogram for
surface-confined redox couple, (B) direct reduction of Subs., (C)
reduction of Subs. at the modified electrode.

method gives a clear qualitative picture, obtaining quantitative
measurements is somewhat involved.

Examples

Reduction of O_2: The $4e^-$ reduction of O_2 to water is a reaction
of great technological relevance in terms of energy conversion via
fuel cells. It is a process that is quite difficult due to the fact
that it is a four-electron process that is also coupled to proton transfers. In addition, competing reactions, such as peroxide formation
can result in a significant voltage loss from the maximum value of
1.23 V.

$$O_2 + 4e^- + 4H^+ \rightarrow 2H_2O \qquad E^\circ = +1.23 \text{ V vs NHE} \qquad (9)$$

$$O_2 + 2e^- + 2H^+ \rightarrow H_2O_2 \qquad E^\circ = +0.68 \text{ V vs NHE} \qquad (10)$$

A number of approaches, including the modification of electrode
surfaces, have been employed in an effort to achieve this transformation. The now-classic work of Anson and Collman [88-90,92,93]
represents one of the most elegant examples of electrocatalysis at
chemically modified electrodes. They employed a family of dicobalt
face-to-face porphyrins in which the length of the link between the
two porphyrins was systematically varied.

Since the generation of peroxide is an undesired reaction pathway, assessments of electrocatalytic efficacy are often made via rotating ring disk electrodes. (Typically, a platinum ring and a graphite disk-configuration is employed). The potential of the ring is set so that any peroxide generated will be detected as an oxidation current. This approach has been followed in numerous studies of electrocatalysis of O_2 reduction.

Murray and coworkers [302] very recently reported a study on the electrocatalysis of O_2 reduction at electrodes modified with electropolymerized films of Co(III) tetra (o-aminophenyl) porphyrin. They noted (Fig. 20) that at low coverage (e.g., 3.4×10^{-10} moles/cm^2) (curve I) a significant amount of peroxide is formed, whereas at high coverage (e.g., 1.6×10^{-9} moles/cm^2) (curve II) the amount of peroxide produced is greatly decreased. This was ascribed to the presence of dimers (or higher aggregates) present at higher coverages, They also report a rate constant of about 1×10^5 $M^{-1}s^{-1}$ for the reduction of O_2, and turnovers in excess of 10,000 at high pH.

Reduction of CO_2. Analogous to the reduction of O_2, the electrocatalytic reduction of CO_2 also presents a great challenge, again due to the need to have multielectronic transformations while avoiding

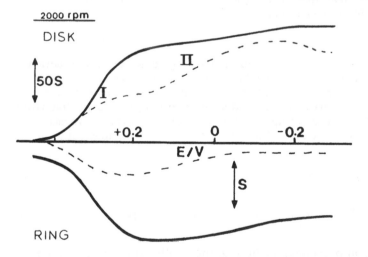

Fig. 20   Disk and ring currents (E_R = +1 V vs Ag/AgCl) vs. disk potential (20 mV/sec) for dioxygen reduction at a 2000 rpm rotating ring-disk electrode coated with 3.4×10^{-10} (I) and 1.6×10^{-9} moles/cm^2 (II) poly(tetra(o-amino) phenyl porphyrin). (From Ref. 302, used with permission.)

high energy intermediates. A number of transition metal complexes have been shown to effect some degree of catalysis in the reduction of CO_2. In addition some of these materials have been employed as modifiers of electrode surfaces.

Murray and Meyer as well as our group [303,215,304,305] have employed electrodes modified with polymeric films of $[Re(CO)_3(v-bpy)Cl]$ for the electrocatalytic reduction of CO_2. The choice of this system originated from solution studies by Lehn and coworkers [306]. Although high catalytic activity was observed (significant currents at -1.4 V), the turnover numbers were modest (ca. 600), presumably due to a dimerization reaction of the catalyst which effectively competes with the electrocatalytic step.

Most recently we [307] have found that the electrodes modified with electropolymerized layers of $[Co(v-terpy)_2]^{2+}$ exhibit a high electrocatalytic activity towards CO_2 reduction (Fig. 21).

B. Stabilization of Small Band Gap Semiconductors

The use of semiconductor electrodes in photoelectrochemical cells represents an efficient and potentially important way of converting solar energy into either electricity or high energy redox products. One of the problems that has plagued this area is the inherent instability of many of these materials under irradiation, especially small band gap non-oxide n-type materials. This is due to the fact that upon absorption of a photon (to generate an electron in the conduction band and a hole in the valence band), the photogenerated hole can react with either a redox system in solution (the desired reaction) or, alternatively, it can oxidize the semiconductor itself. This process of photoanodic dissolution is especially severe for materials such as n-Si and n-GaAs, which are potentially two of the most efficient materials for the conversion of solar energy. The problem, in its simplest form, can be reduced to one of relative rates (i.e., rate of interfacial charge transfer to a redox system in solution vs. dissolution of the electrode), and many investigators have proposed and employed numerous approaches aimed at accelerating interfacial change transfer, thereby stabilizing the semiconductor.

Early on, Wrighton and coworkers [17–19] proposed the modification of the surface of the semiconductor with a polymeric film containing a reversible, outer sphere redox couple, envisioning that this would accelerate interfacial charge transfer and thereby stabilize the semiconductor. An added feature was the fact that the local concentration of redox sites would be of the order of $1\,M$, thereby further enhancing stability. With this purpose in mind, Wrighton and coworkers developed a series of hydrolytically unstable ferrocenyl silanes (Fig. 10) that could be employed in the surface modification of these semiconductor surfaces. As shown in Fig. 22, the use of these materials had a dramatic effect in enhancing the long-term

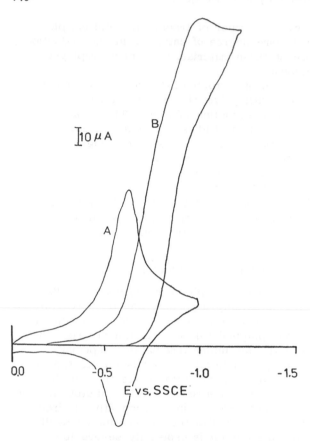

Fig. 21   Cyclic voltammogram for an electrode coated with a polymeric film of $[Co(v\text{-terpy})_2]^{2+}$ under an atmosphere of (A) argon and (B) CO_2.

stability of these electrodes under operation.   These early findings of Wrighton prompted vigorous research by many investigators so that today many redox systems have been incorporated onto a wide variety of semiconductor materials [155,156,308-316].

C.   Electroanalysis

Modified electrodes are ideally suited for the development of analytical techniques, particularly for the determination of transition metal ions. This is due to the inherent sensitivity of the method (recall that a monolayer of metal complex represents about 1×10^{-10} moles/cm^2) and the fact that the interface can be tailored to contain a wide

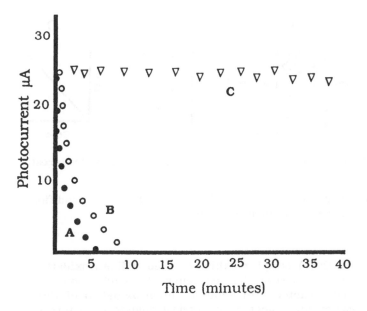

Fig. 22   Stabilization of n-Si in aqueous solution. Plots of photo-current vs. time.   (A) naked Si in contact with 0.1 M NaClO$_4$, (B) same as (A) but in the presence of 4 mM Fe(CN)$_6^{4-}$, (C) same as (B) but electrode modified with (1,1'-ferrocenyl) dichloro silane. (Adapted from Ref. 158.)

range of reactants that can coordinate to a metal center. Thus, the approach is to perform the equivalent of stripping voltammetry in which the preconcentration step is driven by coordination, not electro-deposition. This has the advantage of allowing for unprecedented levels of selectivity and sensitivity.

We recently proposed and demonstrated [317–319] an approach based on the use of functionalized polymer films that contain a ligand (chosen so as to exhibit high affinity and selectivity for a metal ion) as well as an additional redox couple which serve as an internal standard.

The technique depicted in Fig. 23 involves the modification of the electrode, exposure to the analysis solution, and the subsequent electrochemical determination of the metal/ligand complex. From the electrochemical response and a calibration curve, the concentration of the metal ion in solution can be determined.

An approach that we have employed for ligand immobilization has been ion exchange, due to its simplicity. Figure 24 shows the de-termination of Cu(I) with the ligand shown in E incorporated by

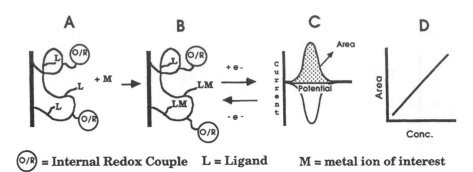

O/R = Internal Redox Couple   L = Ligand   M = metal ion of interest

Fig. 23   Schematic representation of the use of chemically modified electrodes for electroanalysis of metal ions.

ion exchange.   The Cu(I) complex exhibits a well-defined oxidation wave at about +0.35 V and from the peak current a calibration curve (F) can be constructed.   In addition, upon oxidation of Cu(I) to Cu(II), the complex undergoes a geometrical change from tetrahedral to square planar, and due to the steric crowding of the 2,9 dimethyl substituents, dissociates.   However, the ligand remains intact so that the modified electrode can be employed for numerous determinations (A, C, and D).   The analytical applications of modified electrodes is one area that awaits further development, which we and others are currently pursuing [129–133, 320–323].   The recent review by Murray, Ewing and Durst [324] also poses some tantalizing prospects.

D.   Electrorelease of Chemical Reagents

Miller and coworkers [137–140] have demonstrated that it is possible to employ electrochemical techniques to release reagents from polymer-modified electrodes.   The basic approach is to modify the surface of an electrode with a polymer film that contains the desired reagent bound via a linkage that can be electrochemically cleaved.   They have employed a series of polymers which contained neurotransmitters such as dopamine, glutamate, and GABA (γ-aminobutyric acid) bound via an amide linkage.   Upon reduction, the amide bond is cleaved, thus releasing the desired reagent (Fig. 25).   These electrodes had limited capacity since only a fraction of the bound neurotransmitter could be cleaved electrochemically.   To overcome this, Miller and coworkers most recently employed polypyrrole [140] as a matrix within which to incorporate glutamate.   In its oxidized form, polypyrrole is conducting and, in addition, requires charge-compensating

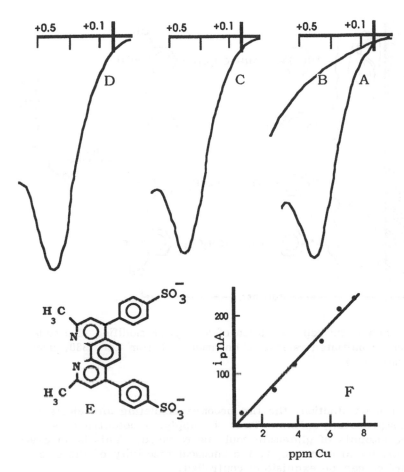

Fig. 24 Determination of Cu^I with electrodes modified with poly-$[Ru(v\text{-}bpy)_3]^{2+}$ and with the ligand shown in E incorporated via ion exchange. (A), (C), and (D) replicate differential pulse voltammograms after exposure of the modified electrode to a (10^{-5} M solution of Cu^I). (B) was made immediately after curve (A), depicting the loss of the copper from the film, once oxidized. (F) is the calibration curve. (Adapted from Ref. 318.)

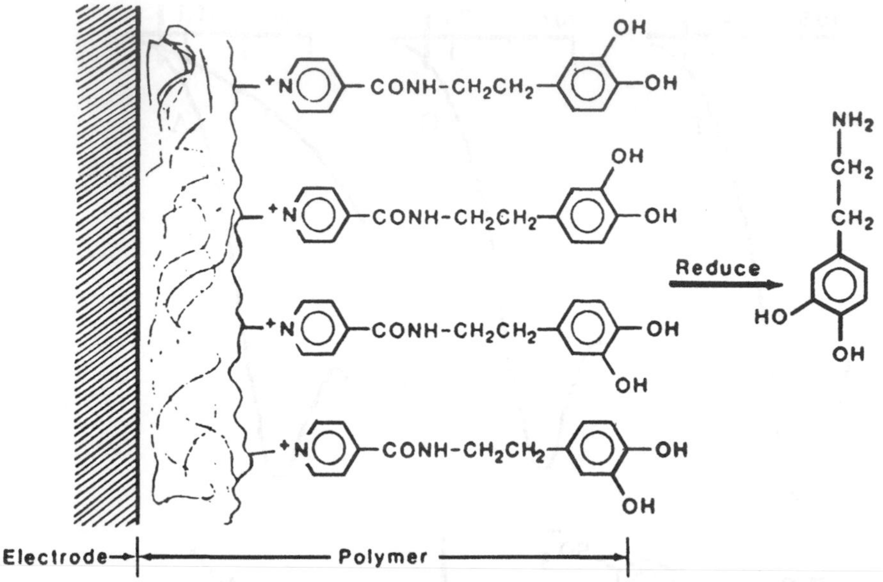

Fig. 25   Schematic representation of a polymer-modified electrode for the electroinitiated release of dopamine. (From Ref. 330, used with permission.)

anions. Upon reduction, the film becomes insulating and ejects the charge-compensating anions. Thus, by applying potential pulses, controlled amounts of glutamate could be released. This is an example of how, by proper design, the chemical reactivity of the electrode surface can be exquisitely controlled.

E. Novel Devices

Novel devices can be prepared by the deposition of structurally segregated and energetically controlled (via the redox potential) films of electroactive materials. Figure 26 depicts a number of such microstructures that have been prepared. The earliest example of this was the so called bilayer electrode, which exhibits bistable switching behavior [24,25,208,213,218]. To prepare such a device, an electrode is coated with two layers of redox-active materials such that the inner layer has two redox processes that straddle the redox process of the outer film. An example of this is presented in Fig. 27; the inner film is a polymeric layer of $[Ru(v\text{-}bpy)_3]^{2+}$ (with redox potentials at about +1.3, −1.35, −1.55, and −1.78 V) and the

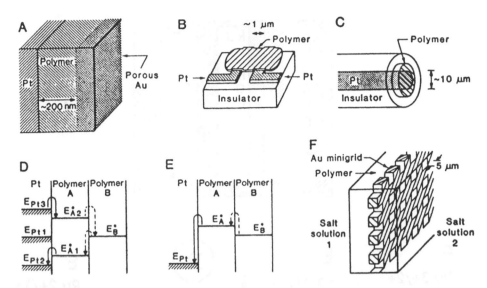

Fig. 26   Microstructures built from electroactive polymers. (A)
sandwich electrode, (B) array electrode, (C) microelectrode, (D)
bilayer electrode type 1, (E) bilayer electrode type 2, and (F) ion-
gate electrode. Except for (A) and (B), which can also be used in
the absence of solvent, an electrolyte solution, reference electrode,
and counter electrode are typically present. (From Ref. 214, used
with permission.)

outer film is a layer of polyvinyl ferrocene (redox potential at about
+0.5 V). The operation of this device can be understood by considering what happens when we vary the applied potential. If we
start with both the $[Ru(bpy)_3]^{2+}$ and ferrocene in their reduced form
and sweep the potential in the positive direction, the direct oxidation
of the ferrocene is not observed because the inner (ruthenium) film
renders it inaccessible. When potentials for the oxidation of the
ruthenium film are reached, both the ruthenium and ferrocene polymers become oxidized. Reversal of the potential in the positive direction reduces the inner ruthenium film back to the 2+ oxidation state.
However, since this species is thermodynamically unable to reduce
the ferricenium sites, the latter are trapped in their oxidized state.
They can be released by scanning in the negative direction (towards
the $Ru^{2+/1+}$ redox couple). The Ru^{1+} species can easily reduce the
ferricenium sites and thus both inner and outer films are reduced
at this potential Thus, this device has rectivying properties (as
depicted in Figure 27c), even though no semiconducting material is
employed.

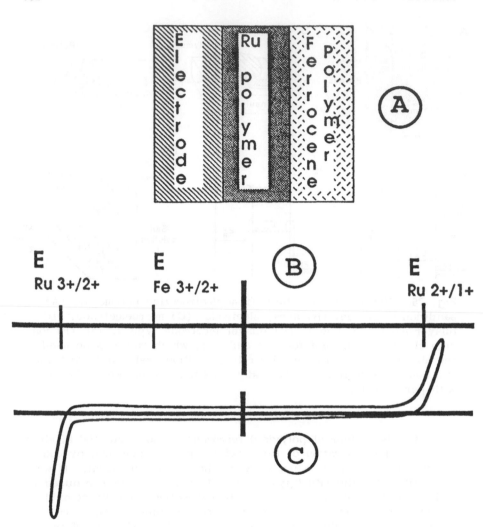

Fig. 27   (A) Schematic representation of a bilayer electrode.   (B) Redox levels of the polymer films.   (C) Current/voltage curve for a bilayer electrode depicting rectifying behavior.

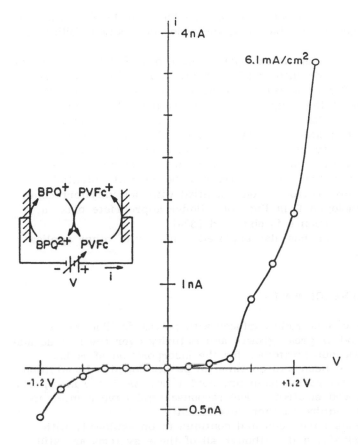

Fig. 28 Two-terminal steady-state current voltage (diode) curve
for electrodes modified with a viologen polymer (BPQ$^+$) and polyvinyl
ferrocene and in which the polymer films are in contact with each
other.   The right-hand portion of the current voltage curve corresponds to the situation in which the negative lead is attached to the
viologen-modified electrode, and the positive lead to the electrode
modified with polyvinyl/ferrocene.   (From Ref. 326, used with permission of the American Chemical Society.)

In addition to bilayer electrodes, Murray has also developed a number of other arrangements based on the redox conduction ability of these films [214].

Another especially elegant example of the types of devices that can be prepared was by Wrighton's [26] group, who employed a microelectrode array (3 electrodes) coated with a polypyrrole film. By taking advantage of the large variation in conductivity of the polypyrrole film with applied potential they were able to prepare a molecular-based transitor. Although the switching times were slow, it demonstrated the types of structures and effects that can be achieved by the control of the chemistry and spatial details. Another example is shown in Fig. 28. It presents the current potential curve for two microelectrodes, one modified with polyvinylferrocene, the other with viologen D in Fig. 10. Under appropriate bias, a typical diode-like response is observed [326].

This research group has also explored a number of other applications [325-330].

VI. FUTURE DEVELOPMENTS

The modification of electrode surfaces with polymeric films has experienced a period of great growth and maturity over the last decade. As a result, numerous strategies for the incorporation of redox-active species have been and continue to be developed. In addition, the theoretical descriptions have provided a framework for interpretation of the observed electrochemical responses and have pointed to those areas that require further development.

Many applications have been and continue to be explored, with emphasis on catalysis, and, although all of these systems are still limited to the confines of research laboratories, I am confident that large scale applications will emerge.

The preparation of structured interfaces of well-defined and deliberately designed character offer some of the most tantalizing future prospects, especially with regard to the development of chemical sensors.

In the process of development, the field has attracted numerous practitioners from other areas. As this trend continues new methods and applications will continue to emerge as we try to dictate the properties of electrochemical interfaces by deliberate modification.

ACKNOWLEDGMENTS

Our work was generously supported by the National Science Foundation, Dow Chemical Co., C. S. Johnson & Son, Inc., Honeywell Inc., Eastman Kodak Co. and the Materials Science Center at Cornell University.

REFERENCES

1. R. W. Murray, *Electroanalytical Chemistry*, Vol. 13, A. J. Bard, ed., Marcel Dekker, New York, 1984, p. 191.
2. R. W. Murray, *Accts. Chem. Res.*, *13*:135 (1980).
3. L. R. Faulkner, *Chem. Eng. News*, Feb. 27, 1982, p. 28.
4. K. D. Snell and A. G. Keenan, *Chem. Soc. Rev.*, *8*:259 (1979).
5. W. J. Alberry and A. R. Hillman, *Ann. Rev. C. Royal Soc. Chem. London*, 1981, p. 377.
6. M. Fujihira, *Topics in Organic Electrochemistry*, A. J. Fry and W. R. Britton, eds., Plenum, New York, 1986, p. 255.
7. J. S. Miller, ed., *Chemically Modified Surfaces in Catalysis and Electrocatalysis*, ACS Symp. Series No. 192, American Chem. Soc., Wash. D.C., 1982.
8. R. W. Murray, *Ann. Rev. Mat. Sci.*, *14*:145 (1984).
9. H. G. Cassidy and K. A. Kun, *Oxidation Reduction Polymers (Redox Polymers)*, Wiley Interscience, N.Y., 1965.
10. F. C. Anson, *Accts. Chem. Res.*, *8*:400 (1975).
11. R. F. Lane and A. T. Hubbard, *J. Phys. Chem.*, *77*:1401 (1973).
12. R. F. Lane and A. T. Hubbard, *J. Phys. Chem.*, *77*:1411 (1973).
13. B. F. Watkins, J. R. Behling, E. Kariv, and L. L. Miller, *J. Am. Chem. Soc.*, *97*:3549 (1975).
14. P. R. Moses, L. M. Wier, and R. W. Murray, *Anal. Chem.*, *47*: 1882 (1975).
15. J. R. Lenhard and R. W. Murray, *J. Electroanal. Chem.* *78*: 195 (1977).
16. J. R. Lenhard, R. Rocklin, H. Abruña, K. Willman, K. Kuo, R. Nowak, and R. W. Murray, *J. Am. Chem. Soc.*, *100*:5213 (1978).
17. M. S. Wrighton, R. G. Austin, A. B. Bocarsly, J. M. Bolts, O. Haas, K. D. Legg, L. Nadjo, and M. C. Palazzotto, *J. Electroanal. Chem.*, *87*:429 (1978).
18. M. S. Wrighton, M. C. Palazzotto, A. B. Bocarsly, J. M. Bolts, A. Fischer, and L. Nadjo, *J. Am. Chem. Soc.*, *100*: 7264 (1978).
19a. M. S. Wrighton, *Accts. Chem. Res.*, *12*:303 (1979).
19b. K. A. Daube, D. J. Harrison, T. E. Mallouk, A. J. Ricco, S. Chao, M. S. Wrighton, W. A. Hendrickson, and A. J. Drube, *J. Photochem.*, *29*:71 (1985).
20a. L. L. Miller and M. R. Van de Mark, *J. Am. Chem. Soc.*, *100*: 3223 (1978).
20b. L. L. Miller and M. R. Van de Mark, *J. Am. Chem. Soc.*, *100*: 639 (1978).
20c. J. B. Kerr and L. L. Miller, *J. Electroanal. Chem.*, *101*:263 (1979).

20d. C. Degrand and L. L. Miller, *J. Electroanal. Chem.*, *117*:267 (1981).

20e. M. Fukui, C. Degrand, and L. L. Miller, *J. Am. Chem. Soc.*, *104*:28 (1982).

20f. C. Degrand and L. L. Miller, *J. Electroanal. Chem.*, *132*:163 (1982).

21a. N. Oyama and F. C. Anson, *J. Electrochem. Soc.*, *127*:247 (1980).

21b. N. Oyama and F. C. Anson, *J. Electrochem. Soc.*, *127*:249 (1980).

22. I. Rubinstein and A. J. Bard, *J. Am. Chem. Soc.*, *102*:6641 (1980).

23. N. Oyama, T. Shimomura, K. Shigehara, and F. C. Anson, *J. Electroanal. Chem.*, *112*:271 (1980).

24. H. D. Abruña, P. Denisevich, M. Umaña, T. J. Meyer, and R. W. Murray, *J. Am. Chem. Soc.*, *103*:1 (1981).

25. P. Denisevich, K. W. Willman, and R. W. Murray, *J. Am. Chem. Soc.*, *103*:4727 (1981).

26. H. S. White, G. P. Kittlesen, and M. S. Wrighton, *J. Am. Chem. Soc.*, *106*:5375 (1984).

27. P. G. Pickup and R. W. Murray, *J. Electrochem. Soc.*, *131*: 833 (1984).

28. T. A. Skotheim, ed., *Handbook of Conducting Polymers*, Marcel Dekker, New York, 1986.

29. A. J. Bard and L. R Faulkner, *Electrochemical Methods*, John Wiley & Sons, Inc., New York, 1980, Chapter 6.

30. E. Laviron, *J. Electroanal. Chem.*, *39*:1 (1972)

31a. E. Laviron, *J. Electroanal. Chem.*, *52*:395 (1974).

31b. E. Laviron, *J. Electroanal. Chem.*, *100*:263 (1979).

31c. E. Laviron, *J. Electroanal. Chem.*, *105*:25 (1979).

31d. E. Laviron, *J. Electroanal. Chem.*, *105*:35 (1979).

32. B. E. Conway and H. Angerstein-Kozlowska, *Accts. Chem. Res.*, *14*:154 (1981).

33. A. P. Brown and F. C. Anson, *Anal. Chem.*, *49*:1589 (1977).

34. D. F. Smith, K. Willman, K. Kuo, and R. W. Murray, *J. Electroanal. Chem.*, *95*:217 (1979).

35. J. C. Lennox and R. W. Murray, *J. Am. Chem. Soc.*, *100*: 3710 (1978).

36. L. L. Miller and M. R. Van de Mark, *J. Am. Chem. Soc.*, *100*: 3223 (1978).

37. R. Nowak, F. A. Schultz, M. Umaña, H. Abruña, and R. W. Murray, *J. Electroanal. Chem.*, *94*:219 (1978).

38. A. H. Schroeder, F. B. Kaufman, V. Patel, and E. M. Engler, *J. Electroanal. Chem.*, *113*:193 (1980).

39. A. Merz and A. J. Bard, *J. Am. Chem. Soc.*, *100*:3222 (1978).

40.   F. B. Kaufman, A. H. Schroeder, E. M. Engler, S. R. Kramer, and J. Q. Chambers, *J. Am. Chem. Soc.*, *102*:483 (1980).

41.   P. Daum and R. W. Murray, *J. Electroanal. Chem.*, *85*:389 (1981).

42.   S. Nakahama and R. W. Murray, *J. Electroanal. Chem.*, *158*: 303 (1983).

43.   P. J. Peerce and A. J. Bard, *J. Electroanal. Chem.*, *114*:89 (1980).

44.   K. W. Willman, R. D. Rocklin, R. Nowak, K. Kuo, F. A. Schultz, and R. W. Murray, *J. Am. Chem. Soc.*, *102*:7629 (1980).

45.   J. B. Kerr, L. L. Miller, and M. R. Van de Mark, *J. Am. Chem. Soc.*, *102*:3383 (1980).

46.   P. J. Peerce and A. J. Bard, *J. Electroanal. Chem.*, *112*:97 (1980).

47.   K. Itaya and A. J. Bard, *Anal. Chem.*, *50*:1487 (1978).

48.   A. H. Schroeder and F. B. Kaufman, *J. Electroanal. Chem.*, *113*:209 (1980).

49.   G. J. Samuels and T. J. Meyer, *J. Am. Chem. Soc.*, *103*:307 (1981).

50.   J. Leddy, A. J. Bard, J. T. Maloy, and J.-M. Savent, *J. Electroanal. Chem.*, *187*:205 (1985).

51.   P. Joo and J. Q. Chambers, *J. Electrochem. Soc.*, *132*:1345 (1985).

52.   J. Q. Chambers and G. Inzelt, *Anal. Chem.*, *57*:1117 (1985).

53.   G. Inzelt, J. Bacskai, J. Q. Chambers, and R. W. Day, *J. Electroanal. Chem.*, *201*:301 (1986).

54.   P. Daum, J. R. Lenhard, D. R. Rolison, and R. W. Murray, *J. Am. Chem. Soc.*, *102*:4649 (1980).

55.   R. J. Nowak, F. A. Schultz, M. Umaña, R. Lam, and R. W. Murray, *Anal. Chem.*, *52*:315 (1980).

56.   N. Oyama and F. C. Anson, *J. Electrochem. Soc.*, *127*:640 (1980).

57.   P. Daum and R. W. Murray, *J. Electroanal. Chem.*, *103*:289 (1979).

58.   P. Denisevich, H. D. Abruña, C. R. Leidner, T. J. Meyer, and R. W. Murray, *Inorg. Chem.*, *21*:2153 (1982).

59.   J. Facci and R. W. Murray, *J. Phys. Chem.*, *86*:2870 (1981).

60.   J. Facci and R. W. Murray, *J. Electroanal. Chem.*, *124*:339 (1981).

61.   E. Laviron, *J. Electroanal. Chem.*, *112*:1 (1980).

62.   D. A. Buttry and F. C. Anson, *J. Electroanal. Chem.*, *130*: 333 (1981).

63.   D. A. Buttry and F. C. Anson, *J. Am. Chem. Soc.*, *104*: 4824 (1982).

64. N. Oyama, S. Yamaguchi, Y. Nishiki, K. Tokuda, H. Matsuda, and F. C. Anson, *J. Electroanal. Chem.*, *139*:371 (1982).

65. R. J. Mortimer and F. C. Anson, *J. Electroanal. Chem.*, *138*: 325 (1982).

66. F. C. Anson, J.-M. Saveant, and K. Shigehara, *J. Am. Chem. Soc.*, *105*:1096 (1983).

67. T. Ohsaka and F. C. Anson, *J. Phys. Chem.*, *87*:640 (1983).

68. D. A. Buttry and F. C. Anson, *J. Am. Chem. Soc.*, *105*:685 (1983).

69. F. C. Anson, J.-M. Saveant, and K. Shigehara, *J. Phys. Chem.*, *87*:214 (1983).

70. H. S. White, J. Leddy, and A. J. Bard, *J. Am. Chem. Soc.*, *104*:4811 (1982).

71. C. R. Martin, I. Rubinstein, and A. J. Bard, *J. Am.Chem. Soc.*, *104*:4817 (1982).

72. J. Leddy and A. J. Bard, *J. Electroanal. Chem.*, *153*:223 (1983).

73. C. R. Martin and K. A. Dollard, *J. Electroanal. Chem.*, *159*: 127 (1983).

74. N. S. Lewis and M. S. Wrighton, *J. Phys. Chem.*, *88*:2009 (1984).

75. T. J. Lewis, H. S. White, and M. S. Wrighton, *J. Am. Chem. Soc.*, *106*:6947 (1984).

76. R. E. Simon, T. E. Mallouk, K. A. Daube, and M. S. Wrighton, *Inorg. Chem.*, *24*:3119 (1985).

77. D. J. Harrison, K. A. Daube, and M. S. Wrighton, *J. Electroanal. Chem.*, *163*:93 (1984).

78. T. Ohsaka, T. Okajima, and N. Oyama, *J. Electroanal. Chem.*, *215*:191 (1986).

79. N. Oyama, T. Ohsaka, and T. Ushirougouchi, *J. Phys. Chem.*, *88*:5274 (1984).

80. N. Oyama, T. Ohsaka, H. Yamamoto, and M. Kaneko, *J. Phys. Chem.*, *90*:3850 (1986).

81. K. Aoki, K. Tokuda, H. Matsuda, and N. Oyama, *J. Electroanal. Chem.*, *176*:139 (1984).

82. T. Ohsaka, N. Oyama, K. Sato, and H. Matsuda, *J. Electrochem. Soc.*, *132*:1871 (1985).

83a. T. P. Henning, H. S. White, and A. J. Bard, *J. Am. Chem. Soc.*, *103*:3937 (1981).

83b. A. J. Bard, T. P. Henning, and H. S. White, *J. Am. Chem. Soc.*, *104*:5362 (1982).

84. T. P. Henning and A. J. Bard, *J. Electrochem. Soc.*, *130*:613 (1983).

85. M. Majda and L. R. Faulkner, *J. Electroanal. Chem.*, *169*:77, 97 (1984).

86a. H. J. Dahms, *J. Phys. Chem.*, *72*:362 (1968).

86b. I. Ruff, *Electrochim. Acta*, *15*:1059 (1970).

86c. I. Ruff and I. Korosi-Odor, *Inorg. Chem.*, *9*:186 (1970).

86d. I. Ruff and V. Friedrich, *J. Phys. Chem.*, *75*:3297, 3303 (1971).

87a. A. P. Brown and F. C. Anson, *J. Electroanal. Chem.*, *92*: 133 (1978).

87b. A. P. Brown, C. Koval, and F. C. Anson, *J. Electroanal. Chem.*, *72*:379 (1976).

88.   J. P. Collman, M. Marrocco, P. Denisevich, C. Koval, and F. C. Anson, *J. Electroanal. Chem.*, *101*:117 (1979).

89.   J. P. Collman, P. Denisevich, Y. Konai, M. Marrocco, C. Koval, and F. C. Anson, *J. Am. Chem. Soc.*, *102*:6027 (1980).

90.   J. P. Collman, F. C. Anson, S. Bencosme, A. Chong, T. Collins, P. Denisevich, E. Evitt, T. Geiger, J. Ibers, G. Janeson, Y. Konai, C. Koval, M. Meier, R. Oakley, R. Pettman, E. Schmitlou, and J. Sessler, *Organic Synthesis Today and Tomorrow*, B. M. Trost and C. R. Hutchinson, eds., Pergamon Press, N.Y., 1981.

91.   R. R. Durand, Jr. and F. C. Anson, *J. Electroanal. Chem.*, *134*:273 (1982).

92.   J. P. Collman, F. C. Anson, C. E. Barnes, C. S. Bencosme, T. Geiger, E. R. Evitt, R. P. Kreh, K. Meier, and R. B. Pettman, *J. Am. Chem. Soc.*, *105*:2694 (1983).

93.   J. P. Collman, F. C. Anson, C. S. Bencosme, R. R. Durand, Jr., and R. P. Kreh, *J. Am. Chem. Soc.*, *105*:2699 (1983).

94.   R. R. Durand, Jr., J. P. Collman, and F. C. Anson, *J. Electroanal. Chem.*, *105*:271 (1983).

95.   R. R. Durand, Jr., J. P. Collman, and F. C. Anson, *J. Electroanal. Chem.*, *151*:289 (1983).

96.   H.-Y. Liu, I. Abdalmuhdi, C. K. Chang, and F. C. Anson, *J. Phys. Chem.*, *89*:665 (1985).

97a. J. S. Facci, P. A. Falcigno, and J. M. Gold, *Langmuir*, *2*: 732 (1986).

97b. M. Fujihira and T. Araki, *Bull. Chem. Soc. Jpn.*, *59*:2375 (1986).

97c. M. Fujihira and T. Araki, *J. Electroanal. Chem.*, *205*:329 (1986).

97d. H. Daipuku, I. Yoshimura, I. Hirata, K. Aoki, K. Tokuda, and H. Matsuda, *J. Electroanal. Chem.*, *199*:47 (1986).

97e. H. Daipuku, K. Aoki, K. Tokuda, and H. Matsuda, *J. Electroanal. Chem.*, *183*:1 (1985).

97f. M. Fujihira and S. Poosittisak, *J. Electroanal. Chem.*, *199*: 481 (1986).

98.   P. R. Moses and R. W. Murray, *J. Am. Chem. Soc.*, *98*: 7435 (1976).

99.   D. F. Untereker, J. C. Lennox, L. M. Wier, P. R. Moses, and
      R. W. Murray, *J. Electroanal. Chem.*, *81*:309 (1977).

100. P. R. Moses and R. W. Murray, *J. Electroanal. Chem.*, *77*:
      393 (1977).

101. J. R. Lenhard and R. W. Murray, *J. Am. Chem. Soc.*, *100*:
      7870 (1978).

102. K. Kuo, P. R. Moses, J. R. Lenhard, D. C. Green, and
      R. W. Murray, *Anal. Chem.*, *51*:745 (1979).

103. H. Abruña, T. J. Meyer, and R. W. Murray, *Inorg. Chem.*,
      11:3233 (1979).

104. H. O. Finklea, H. Abruña, and R. W. Murray, *Interfacial
      Photoprocesses: Energy Conversion and Synthesis*, M. S.
      Wrighton, ed., Adv. in Chem. Series 184, American Chemical
      Society, Wash. D.C., 1980, p. 253.

105. H. D. Abruña, J. L. Walsh, T. J. Meyer, and R. W. Murray,
      J. Am. Chem. Soc., *102*:3272 (1980).

106. H. D. Abruña, J. L. Walsh, T. J. Meyer, and R. W. Murray,
      Inorg. Chem., *20*:1481 (1981).

107. P. R. Moses, L. M. Wier, J. C. Lennox, H. O. Finklea,
      J. R. Lehnard, and R. W. Murray, *Anal. Chem.*, *50*:576
      (1978).

108. H. O. Finklea and R. W. Murray, *J. Phys. Chem.*, *83*:353
      (1979).

109. A. B. Fischer, M. S. Wrighton, M. Umaña, and R. W.
      Murray, *J. Am. Chem. Soc.*, *101*:3442 (1979).

110. K. W. Willman, E. Greer, and R. W. Murray, *Nouv. J. Chemie.*,
      3:455 (1979).

111a. G. J. Leigh and C. J. Pickett, *J. Chem. Soc. Dalton*, 1977,
      p. 1797.

111b. R. J. Burt, G. J. Leigh, and C. J. Pickett, *J. Chem. Soc.
      Chem. Commun.*, 1976, p. 940.

112a. J. C. Lennox and R. W. Murray, *J. Electroanal. Chem.*, *78*:
      395 (1977).

112b. J. C. Lennox and R. W. Murray, *J. Am. Chem. Soc.*, *100*:
      3710 (1978).

113. R. D. Rocklin and R. W. Murray, *J. Electroanal. Chem.*, *100*:
      271 (1979).

114. C. P. Jester, R. D. Rocklin, and R. W. Murray, *J. Electro-
      chem. Soc.*, *127*:1979 (1980).

115. C. M. Elliott and C. A. Marresse, *J. Electroanal. Chem.*, *119*:
      395 (1981).

116. N. Oyama, K. B. Yap, and F. C. Anson, *J. Electroanal.
      Chem.*, *100*:233 (1979).

117. C. A. Koval and F. C. Anson, *Anal. Chem.*, *50*:233 (1978).

118. N. Oyama and F. C. Anson, *J. Am. Chem. Soc.*, *101*:1634
      (1979).

119. A. W. C. Lin, P. Yeh, A. M. Yacynych, and T. Kuwana, *J. Electroanal. Chem.*, *84*:411 (1971).
120. A. M. Yacynych and T. Kuwana, *Anal. Chem.*, *50*:640 (1978).
121. D. C. S. Tse, T. Kuwana, and G. P. Royer, *J. Electroanal. Chem.*, *98*:345 (1979).
122. M. F. Dautartas, J. F. Evans, and T. Kuwana, *Anal. Chem.*, *51*:104 (1979).
123. M. A. Fox, F. J. Nobs, and T. A. Voynick, *J. Am. Chem. Soc.*, *102*:4029 (1980).
124a. R. M. Ianello and A. M. Yacynych, *Anal. Chim. Acta*, *131*: 123 (1981).
124b. R. M. Ianello and A. M. Yacynych, *Anal. Chem.*, *53*:2090 (1981).
125. T. Kuwana and W. G. French, *Anal. Chem.*, *36*:241 (1964).
126. F. A. Schultz and T. Kuwana, *J. Electroanal. Chem.*, *10*:95 (1965).
127. D. Baver and M. P. Gaillochet, *Electrochim. Acta*, *19*:592 (1974).
128. M. Lamanche and D. Bauer, *J. Electroanal. Chem.*, *79*:359 (1977).
129a. R. Ravichandron and R. P. Baldwin, *J. Electroanal. Chem.*, *126*:293 (1981).
129b. R. Ravichandron and R. P. Baldwin, *Anal. Chem.*, *55*:1586 (1983).
130a. W. Kutner, T. J. Meyer, and R. W. Murray, *J. Electroanal. Chem.*, *105*:375 (1985).
130b. E. S. Takeuchi and R. W. Murray, *J. Electroanal. Chem.*, *188*:49 (1985).
131. G. T. Cheek and R. F. Nelson, *Anal. Lett.*, *A11*:393 (1978).
132. T. Yao and S. Muscha, *Anal. Chim. Acta*, *110*:203 (1979).
133. R. P. Baldwin, J. C. Christensen, and L. Kryger, *Anal. Chem.*, *58*:1790 (1986).
134. J. B. Kerr, L. L. Miller, and M. R. Van de Mark, *J. Am. Chem. Soc.*, *102*:3383 (1980).
135. L. L. Miller, B. Zinger, and C. Degrand, *J. Electroanal. Chem.*, *178*:87 (1984).
136. C. Degrand and L. L. Miller, *J. Am. Chem. Soc.*, *102*:5728 (1980).
137. L. L. Miller, A. N. K. Lau, and E. K. Miller, *J. Am. Chem. Soc.*, *104*:5242 (1982).
138. A. N. K. Lau and L. L. Miller, *J. Am. Chem. Soc.*, *105*: 5271 (1983).
139. A. N. K. Lau, L. L. Miller, and B. Zinger, *J. Am. Chem. Soc.*, *105*:5278 (1983).
140. L. L. Miller and B. Zinger, *J. Am. Chem. Soc.*, *106*:6861 (1984).

141a.   J. Q. Chambers, F. B. Kaufman, and K. H. Nichols, *J. Electroanal. Chem.*, *142*:277 (1982).

141b.   G. Inzelt, J. Q. Chambers, and F. B. Kaufman, *J. Electroanal. Chem.*, *159*:443 (1983).

142.   A. H. Schroeder and F. B. Kaufman, *J. Electroanal. Chem.*, *113*:209 (1980).

143.   F. B. Kaufman, *IBM J. Res. Develop.*, *25*:303 (1981).

144.   F. B. Kaufman, A. H. Schroeder, V. V. Patel, and K. H. Nichols, *J. Electroanal. Chem.*, *132*:151 (1982).

145.   F. B. Kaufman and E. M. Engler, *J. Am. Chem. Soc.*, *101*:547 (1979).

146.   R. W. Day, G. Inzelt, J. F. Kinstle, and J. Q. Chambers, *J. Am. Chem. Soc.*, *104*:6804 (1982).

147.   G. Inzelt, R. W. Day, J. F. Kinstle, and J. Q. Chambers, *J. Phys. Chem.*, *87*:4592 (1983).

148.   G. Inzelt, R. W. Day, J. F. Kinstle, and J. Q. Chambers, *J. Electroanal. Chem.*, *161*:147 (1984).

149.   G. Inzelt, J. Q. Chambers, J. F. Kinstle, R. W. Day, *J. Am. Chem. Soc.*, *106*:3396 (1984).

150.   G. Inzelt, J. Q. Chambers, J. F. Kinstle, R. W. Day, and M. A. Lange, *Anal. Chem.*, *56*:301 (1984).

151.   R. W. Day, H. Karimi, C. V. Francis, J. F. Kinstle, and J. Q. Chambers, *J. Polym. Sci. Chem. Edit.*, *24*:645 (1986).

152.   G. Inzelt, J. Bacskai, J. Q. Chambers, and R. W. Day, *J. Electroanal. Chem.*, *201*:301 (1986).

153.   B. L. Funt and P. M. Hoang, *J. Electroanal. Chem.*, *154*:229 (1983).

154.   C. Degrand and E. Laviron, *J. Electroanal. Chem.*, *117*:283 (1981).

155.   H. D. Abruña and A. J. Bard, *J. Am. Chem. Soc.*, *103*:6898 (1981).

156.   M. D. Rosenblum and N. S. Lewis, *J. Phys. Chem.*, *88*:3103 (1984).

157a.   M. S. Wrighton, R. Austin, and A. B. Palazzotto, *J. Am. Chem. Soc.*, *100*:1602 (1978).

157b.   J. M. Bolts and M. S. Wrighton, *J. Am. Chem. Soc.*, *100*:5257 (1978).

158.   J. M. Bolts, A. B. Bocarsly, M. C. Palazzotto, E. G. Walton, N. S. Lewis, and M. S. Wrighton, *J. Am. Chem. Soc.*, *101*:1378 (1979).

159.   J. M. Bolts and M. S. Wrighton, *J. Am. Chem. Soc.*, *101*:6179 (1979).

160.   A. B. Bocarsly, E. G. Walton, M. G. Bradley, and M. S. Wrighton, *J. Electroanal. Chem.*, *100*:283 (1979).

161.   A. B. Bocarsly, E. G. Walton, and M. S. Wrighton, *J. Am. Chem. Soc.*, *102*:3390 (1980).

162. N. S. Lewis, A. B. Bocarsly, and M. S. Wrighton, *J. Phys. Chem.*, *84*:2033 (1980).

163. D. C. Bookbinder, J. A. Bruce, R. N. Dominey, N. S. Lewis, and M. S. Wrighton, *Proc. Nat. Acad. Sci. USA*, 77:6280 (1980).

164. D. C. Bookbinder and M. S. Wrighton, *J. Am. Chem. Soc.*, *102*:5123 (1980).

165. N. S. Lewis and M. S. Wrighton, *Science 211*:944 (1981).

166. J. A. Bruce and M. S. Wrighton, *J. Am. Chem. Soc.*, *104*: 74 (1982).

167. R. N. Dominey, N. S. Lewis, J. A. Bruce, D. C. Bookbinder, and M. S. Wrighton, *J. Am. Chem. Soc.*, *104*:467 (1982).

168. R. N. Dominey, T. J. Lewis, and M. S. Wrighton, *J. Phys. Chem.*, *87*:5345 (1983).

169. D. C. Bookbinder and M. S. Wrighton, *J. Electrochem. Soc.*, *130*:1080 (1983).

170. D. J. Harrison and M. S. Wrighton, *J. Am. Chem. Soc.*, *106*: 3932 (1984).

171. C. J. Stadler, S. Chao, and M. S. Wrighton, *J. Am. Chem. Soc.*, *106*:3673 (1984).

172. J.-F. Andree and M. S. Wrighton, *Inorg. Chem.*, *24*:4288 (1985).

173a. G. S. Calabrese, R. M. Buchanan, and M. S. Wrighton, *J. Am. Chem. Soc.*, *105*:5594 (1983).

173b. G. S. Calabrese, R. M. Buchanan, and M. S. Wrighton, *J. Am. Chem. Soc.*, *104*:5786 (1982).

174. R. M. Buchanan, G. S. Calabrese, T. J. Sobieralski, and M. S. Wrighton, *J. Electroanal. Chem.*, *153*:129 (1983).

175. R. A. Simon, T. E. Mallouk, K. A. Daube, and M. S. Wrighton, *Inorg. Chem.*, *24*:3119 (1985).

176. K. A. Daube, D. J. Harrison, T. E. Mallouk, A. J. Ricco, S. Chao, and M. S. Wrighton, *J. Photochem.*, *29*:71 (1985).

177. T. J. Lewis, H. S. White, and M. S. Wrighton, *J. Am. Chem. Soc.*, *106*:6947 (1984).

178. N. S. Lewis and M. S. Wrighton, *Science*, *211*:944 (1981).

179. J. A. Bruce and M. S. Wrighton, *Israel J. Chem.*, *22*:184 (1982).

180. N. S. Lewis and M. S. Wrighton, *J. Phys. Chem.*, *88*:2009 (1984).

181a. C. M. Elliott and E. J. Hersenhart, *J. Am. Chem. Soc.*, *104*: 7519 (1982).

181b. C. M. Elliott, J. G. Redepenning, and E. M. Balk, *J. Am. Chem. Soc.*, *107*:8302 (1985).

181c. C. M. Elliott and J. C. Redepenning, *J. Electroanal. Chem.*, *197*:219 (1986).

182a. W. J. Alberry, A. W. Foulds, K. J. Hall, and A. R. Hillman, *J. Electrochem. Soc.*, *127*:654 (1980).

182b. W. J. Alberry, W. R. Bowen, F. S. Fisher, A. W. Foulds, K. J. Hall, A. R. Hillman, R. G. Egdell, and A. F. Orchard, *J. Electroanal. Chem.*, *107*:37 (1980).

183.   O. Haas, H.-R. Zumbrunnen, *Helv. Chim. Acta*, *64*:855 (1981).

184.   J.-C. Moutet, *J. Electroanal. Chem.*, *161*:181 (1984).

185.   M.-C. Pham, A. Hachemi, and J.-E. Dubois, *J. Electroanal. Chem.*, *161*:199 (1984).

186.   A. Torstensson and L. Gorton, *J. Electroanal. Chem.*, *130*:199 (1981).

187.   G. Bidan, A. Deronzier, and J.-C. Moutet, *J. Chem. Soc., Chem. Commun.*, 1984, p. 1185.

188.   K. W. Willman and R. W. Murray, *J. Electroanal. Chem.*, *133*:211 (1982).

189.   C. M. Elliott and W. S. Martin, *J. Electroanal. Chem.*, *137*:377 (1982).

190.   N. Oyama and F. C. Anson, *J. Am. Chem. Soc.*, *101*:739 (1979).

191.   N. Oyama and F.C. Anson, *J. Am. Chem. Soc.*, *101*:3450 (1979).

192.   N. S. Scott, N. Oyama, and F. C. Anson, *J. Electroanal. Chem.*, *110*:303 (1980).

193.   G.-X. Wan, K. Shigehara, E. Tsuchida, and F. C. Anson, *J. Electroanal. Chem.*, *179*:239 (1984).

194.   J. Calvert and T. J. Meyer, *Inorg. Chem.*, *20*:27 (1981).

195.   G. J. Samuels and T. J. Meyer, *J. Am. Chem. Soc.*, *103*:307 (1981).

196.   J. M. Calvert, B. P. Sullivan, and T. J. Meyer, *Chemical Modification of Surfaces*, J. S. Miller, ed., Adv. in Chem. Series #192, American Chemical Society, Wash. D.C., 1982, p. 159.

197.   C. D. Ellis and T. J. Meyer, *Inorg. Chem.*, *23*:1748 (1984).

198a. J. T. Hupp, J. P. Otruba, S. J. Paurs, and T. J. Meyer, *J. Electroanal. Chem.*, *190*:287 (1985).

198b. T. D. Westmoreland, J. M. Calvert, R. W. Murray, and T. J. Meyer, *J. Chem. Soc. Chem. Commun.*, 1983, p. 65.

199.   L. D. Margerum, R. W. Murray, and T. J. Meyer, *J. Phys. Chem.*, *90*:728 (1986).

200.   O. Haas and J. G. Vos, *J. Electroanal. Chem.*, *113*:139 (1980).

201.   O. Haas, M. Kriens, and J. G. Vos, *J. Am. Chem. Soc.*, *103*:1318 (1981).

202a. O. Haas, H. R. Zumbrunnen, and J. G. Vos, *Electrochim. Acta*, *30*:1551 (1985).

202b. C. P. Andrieux, O. Haas and J.-M. Saveant, *J. Am. Chem. Soc.*, *108*:8175 (1986).

203. S. Geraty and J. G. Vos, *J. Electroanal. Chem.*, *176*:389 (1984).

204a. E. I. Becker, M. Tutsui, eds., *Organometallic Reactions and Synthesis*, Vol. 6, Plenum, N.Y., 1977.

204b. C. E. Carraher, J. E. Sheats, and C. U. Pittman, eds., *Organometallic Polymers*, Academic Press, N.Y., 1978.

204c. A. Akelah and D. C. Sherrington, *Chem. Rev.*, *81*:557 (1981).

204d. P. Hodge and D. C. Sherrington, eds., *Polymer Supported Reactions in Organic Synthesis*, Wiley, N.Y., 1980.

204e. W. T. Ford, ed., *Polymeric Reagents and Catalysis*, ACS Symposium Series 308, American Chemical Society, Washington, D.C., 1986.

205a. J. M. Calvert, R. H. Schmehl, B. P. Sullivan, J. S. Facci, T. J. Meyer, and R. W. Murray, *Inorg. Chem.*, *22*:2151 (1983).

205b. T. F. Guarr and F. C. Anson, *J. Phys. Chem.*, *91*:4037 (1987).

206. T. Ikeda, R. Schmehl, P. Denisevich, K. Willman, and R. W. Murray, *J. Am. Chem. Soc.*, *104*:2683 (1982).

207a. T. Ikeda, C. R. Leidner, and R. W. Murray, *J. Am. Chem. Soc.*, *103*:7422 (1981).

207b. T. Ikeda, C. R. Leidner, and R. W. Murray, *Rev. Polarog. Japan*, *27*:54 (1981).

208. P. Denisevich, H. D. Abruña, C. R. Leidner, T. J. Meyer, and R. W. Murray, *Inorg. Chem.*, *21*:2153 (1982).

209. T. Ikeda, C. R. Leidner, and R. W. Murray, *J. Electroanal. Chem.*, *138*:343 (1982).

210a. C. R. Leidner and R. W. Murray, *J. Am. Chem. Soc.*, *106*: 1606 (1984).

210b. C. R. Leidner and R. W. Murray, *J. Am. Chem. Soc.*, *107*: 551 (1985).

210c. J. S. Facci, R. H. Schmehl, and R. W. Murray, *J. Am. Chem. Soc.*, *104*:4959 (1982).

211. B. J. Feldman, A. E. Ewing, and R. W. Murray, *J. Electroanal. Chem.*, *194*:63 (1985).

212. H. D. Abruña, J. M. Calvert, C. D. Ellis, T. J. Meyer, R. W. Murray, B. P. Sullivan, and J. L. Walsh, *Chemical Modification of Surfaces*, J. S. Miller, ed., Advances in Chemistry Series No. 192, American Chemical Society, Wash. D.C., 1982, p. 133.

213. P. G. Pickup, W. Kutner, C. R. Leidner, and R. W. Murray, *J. Am. Chem. Soc.*, *106*:1991 (1984).

214. C. E. D. Chidsey and R. W. Murray, *Science*, *231*:25 (1986).

215. T. R. O'Toole, L. D. Margerum, T. D. Westmoreland, W. J. Vining, R. W. Murray, and T. J. Meyer, *J. Chem. Soc. Chem. Comm.*, 1985, p. 1416.

216a. S. Kanda and R. W. Murray, *Bull. Chem. Soc. Japan, 58*:
       3010 (1985).

216b. L. D. Margerum, T. J. Meyer, and R. W. Murray, *J. Electro-
       anal. Chem., 149*:279 (1983).

217.   W. J. Vining, N. A. Surridge, and T. J. Meyer, *J. Phys.
       Chem., 90*:2281 (1986).

218a. P. G. Pickup, C. R. Leidner, P. Denisevich, and R. W.
       Murray, *J. Electroanal. Chem., 164*:39 (1984).

218b. C. R. Leidner, P. Denisevich, K. W. Willman, and R. W.
       Murray, *J. Electroanal. Chem., 164*:63 (1984).

219.   J. M. Calvert, D. L. Peebles, and R. J. Nowak, *Inorg. Chem.,
       24*:3111 (1985).

220.   K. T. Potts, D. Usifer, A. Guadalupe, and H. D. Abruña,
       J. Am. Chem. Soc., 109:3961 (1987).

221.   P. G. Pickup and R. A. Osteryoung, *J. Electrochem. Soc.,
       130*:1965 (1983).

222.   C. D. Ellis, L. D. Margerum, R. W. Murray, and T. J.
       Meyer, *Inorg. Chem., 22*:1283 (1983).

223a. K. A. Macor and T. G. Spiro, *J. Am. Chem. Soc., 105*:5601
       (1983).

223b. K. A. Macor and T. G. Spiro, *J. Electroanal. Chem., 163*:
       223 (1984).

223c. R. M. Kellet and T. G. Spiro, *Inorg. Chem., 24*:2378
       (1985).

224.   B. A. White and R. W. Murray, *J. Electroanal. Chem., 189*:
       345 (1985).

225a. S. Cosneir, A. Deronzier, and J.-C. Moutet, *J. Electroanal.
       Chem., 193*:193 (1985).

225b. S. Cosnier, A. Deronzier, and J.-C. Moutet, *J. Phys. Chem.,
       89*:4895 (1985).

225c. S. Cosneir, A. Deronzier, and J.-C. Moutet, *J. Electroanal.
       Chem., 207*:315 (1986).

226.   J. G. Eaves, H. S. Munro, and D. Parker, *J. Chem. Soc.
       Chem. Commun.*, 1985, p. 684.

227.   G. Schiavon, G. Zotti, and G. Bontempelli, *J. Electroanal.
       Chem., 161*:323 (1984).

228.   M. Yamana, R. Darby, H. P. Dhar, and R. E. White, *J.
       Electroanal. Chem., 152*:261 (1983).

229.   J. Asseraf, F. Bedioui, O. Reges, Y. Robin,F. Devynck,
       and C. Bed-Charreton, *J. Electroanal. Chem., 170*:255 (1983).

230.   N. Oyama and F. C. Anson, *Anal. Chem., 52*:1192 (1980).

231a. N. Oyama, K. Sato, and F. C. Anson, *J. Electroanal. Chem.,
       115*:149 (1980).

231b. N. Oyama, T. Ohsaka, and T. Ushirogouchi, *J. Phys. Chem.,
       88*:5274 (1984).

232.   N. Oyama, K. Sato, and H. Matsuda, *J. Electroanal. Chem.,
       115*:149 (1980).

233.   H. R. Zumbrunnen and F. C. Anson, *J. Electroanal. Chem.*, *152*:111 (1983).

234a.   K. Doblhofer, H. Braun, and W. Storck, *J. Electrochem. Soc.*, *130*:807 (1983).

234b.   H. Braun, F. Decker, K. Doblhofer, and H. Sotobayashi, *Ber. Bunsenges. Phys. Chem.*, *88*:345 (1984).

234c.   K. Doblhofer, H. Braun, and R. Lange, *J. Electroanal. Chem.* *206*:93 (1986).

234d.   K. Niwa and K. Doblhofer, *Electrochim. Acta*, *31*:553 (1986).

235.   E. S. De Castro, E. W. Huber, D. Villaroel, C. Galiatsatos, J. E. Mark, W. E. Heineman, and P. T. Murray, *Anal. Chem.*, *59*:134 (1987).

236.   F. C. Anson, T. Ohsaka, and J.-M. Saveant, *J. Am. Chem. Soc.*, *105*:4883 (1983).

237.   K. Sumi and F. C. Anson, *J. Phys. Chem.*, *90*:3845 (1986).

238a.   T. Ohsaka, T. Okajima, and N. Oyama, *J. Electroanal. Chem.*, *215*:191 (1986).

238b.   N. Oyama, T. Ohsaka, and T. Okajima, *Anal. Chem.*, *58*:981 (1986).

239.   B. R. Shaw, G. P. Haight, Jr., and L. R. Faulkner, *J. Electroanal. Chem.*, *140*:147 (1982).

240.   W. J. Vining and T. J. Meyer, *J. Electroanal. Chem.*, *195*: 183 (1985).

241.   W. J. Vining and T. J. Meyer, *Inorg. Chem.*, *25*:2023 (1986).

242.   E. I. DuPont de Nemours, Inc.

243.   A. Eisenberg and H. L. Yeager, eds., *Perfluoro Sulfonate Ionomer Membranes*, A.C.S. Symposium Series 180, American Chemical Society, Washington, D.C., 1982.

244.   M. N. Szentirmay and C. R. Martin, *Anal. Chem.*, *56*:1898 (1984).

245a.   C. R. Martin, *Anal. Chem.*, *54*:1639 (1982).

245b.   R. B. Moore III and C. R. Martin, *Anal. Chem.*, *58*:2569 (1986).

246.   D. A. Buttry and F. C. Anson, *J. Am. Chem. Soc.*, *104*: 4824 (1982).

247.   H. Y. Liu and F. C. Anson, *J. Electroanal. Chem.*, *158*:181 (1983).

248.   D. A. Buttry and F. C. Anson, *J. Am. Chem. Soc.*, *106*:59 (1984).

249.   D. A. Buttry, J.-M. Saveant and F. C. Anson, *J. Phys. Chem.*, *88*:3086 (1984).

250.   R. C. McHatton and F. C. Anson, *Inorg. Chem.*, *23*:3935 (1984).

251.   K. Shigehara, E. Tsuchida and F. C. Anson, *J. Electroanal. Chem.*, *175*:291 (1984).

252.   F. C. Anson, J.-M. Saveant, and Y.-M. Tsou, *J. Electroanal. Chem.*, *178*:113 (1984).

253. F. C. Anson, C.-L. Ni, and J.-M. Saveant, *J. Am. Chem. Soc.*, *107*:3442 (1985).

254. Y.-M. Tsou and F. C. Anson, *J. Phys. Chem.*, *89*:3818 (1985).

255. I. Rubinstein and A. J. Bard, *Anal. Chem.*, *53*:102 (1981).

256. M. Krishnan, J. R. White, M. A. Fox, and A. J. Bard, *J. Am. Chem. Soc.*, *105*:7002 (1983).

257. A. Mau, C. B. Huang, N. Kakuta, A. Campion, M. A. Fox, J. M. White, S. E. Webber, and A. J. Bard, *J. Am. Chem. Soc.*, *106*:6537 (1984).

258. J. G. Gaudiello, P. K. Ghosh, and A. J. Bard, *J. Am. Chem. Soc.*, *107*:3027 (1985).

259. A. E. Kaifer and A. J. Bard, *J. Phys. Chem.*, *90*:868 (1986).

260. J. Leddy and A. J. Bard, *J. Electroanal. Chem.*, *189*:203 (1985).

261. C. R. Martin and K. A. Dollard, *J. Electroanal. Chem.*, *159*:127 (1983).

262. G. Nagy, G. A. Gerhardt, A. F. Oke, M. E. Rice, R. N. Adams, R. B. Moore, III, M. N. Szentirmay, and C. R. Martin, *J. Electroanal. Chem.*, *188*:85 (1985).

263. C. R. Martin, *J. Chem. Soc. Farad. Trans. 1*, *82*:1051 (1986).

264. R. M. Penner and C. R. Martin, *J. Electrochem. Soc.*, *132*:514 (1985).

265. J. Redepenning and F. C. Anson, *J. Phys. Chem.*, *90*:6227 (1986).

266a. C. M. Lieber, M. H. Schmidt, and N. S. Lewis, *J. Am. Chem. Soc.*, *108*:6103 (1986).

266b. C. M. Lieber, M. H. Schmidt, and N. S. Lewis, *J. Phys. Chem.*, *90*:1002 (1986).

267. M. Umaña, P. Denisevich, D. R. Rolison, S. Nakahama, and R. W. Murray, *Anal. Chem.*, *53*:1170 (1981).

268a. M. F. Dautartas and J. F. Evans, *J. Electroanal. Chem.*, *109*:301 (1980).

268b. M. F. Dautartas, K. R. Mann, and J. F. Evans, *J. Electroanal. Chem.*, *110*:379 (1980).

269. J. Facci and R. W. Murray, *Anal. Chem.*, *54*:772 (1982).

270. For a recent review see K. Itaya, I. Uchida, and V. D. Neff, *Accts. Chem. Res.*, *19*:162 (1986).

271. V. D. Neff, *J. Electrochem. Soc.*, *125*:886 (1978).

272. A. B. Bocarsly and S. Sinha, *J. Electroanal. Chem.*, *137*:157 (1982).

273. A. B. Bocarsly and S. Sinha, *J. Electroanal. Chem.*, *140*:167 (1982).

274. A. B. Bocarsly, S. A. Galvin, and S. Sinha, *J. Electrochem. Soc.*, *130*:1319 (1983).

275.   B. D. Humphrey, S. Sinha, and A. B. Bocarsly, *J. Phys. Chem.*, *88*:736 (1984).

276.   S. Sinha, B. D. Humphrey, E. Fu, and A. B. Bocarsly, *J. Electroanal. Chem.*, *162*:351 (1984).

277.   S. Sinha, B. D. Humphrey, and A. B. Bocarsly, *Inorg. Chem.*, *23*:203 (1984).

278.   L. J. Amos, M. H. Schmidt, S. Sinha, and A. B. Bocarsly, *Langmuir*, *2*:561 (1986).

279.   D. W. DeBerry and A. Viehbeck, *J. Electrochem. Soc.*, *130*: 249 (1983).

280a. L. M. Siperko and T. Kuwana, *J. Electrochem. Soc.*, *130*:396 (1983).

280b. L. M. Siperko and T. Kuwana, *J. Electrochem. Soc.*, *133*: 2440 (1986).

281.   K. Itaya, N. Shoji, and I. Uchida, *J. Am. Chem. Soc.*, *106*: 3423 (1984).

282.   T. Ikeshoji, *J. Electrochem. Soc.*, *133*:2108 (1986).

283.   D. Shaojun and L. Fengbin, *J. Electroanal. Chem.*, *210*:31 (1986).

284.   R. J. Mortimer and D. Rosseinsky, *J. Electroanal. Chem.*, *151*:133 (1983).

285.   K. Ogura and M. Takagi, *J. Electroanal. Chem.*, *206*:209 (1986).

286.   C. J. Miller and M. Majda, *J. Am. Chem. Soc.*, *107*:1419 (1985).

287.   C. J. Miller and M. Majda, *J. Electroanal. Chem.*, *207*:49 (1986).

288.   C. J. Miller and M. Majda, *J. Am. Chem. Soc.*, *108*:3118 (1986).

289.   A. J. Bard and P. K. Ghosh, *J. Am. Chem. Soc.*, *105*:5691 (1983).

290.   P. K. Ghosh, A. W.-H. Mau, and A. J. Bard, *J. Electroanal. Chem.*, *169*:315 (1984).

291.   D. Ege, P. K. Ghosh, J. R. White, J.-F. Equey, and A. J. Bard, *J. Am. Chem. Soc.*, *107*:5644 (1985).

292.   J. R. White and A. J. Bard, *J. Electroanal. Chem.*, *197*:233 (1986).

293.   K. Itaya and A. J. Bard, *J. Phys. Chem.*, *89*:5565 (1985).

294.   W. E. Rudzinski and A. J. Bard, *J. Electroanal. Chem.*, *199*: 323 (1986).

295.   H. Y. Liu and F. C. Anson, *J. Electroanal. Chem.*, *184*:411 (1985).

296.   C. G. Murray, R. J. Nowak, and D. R. Rolison, *J. Electroanal. Chem.*, *164*:205 (1984).

297.   T. J. Meyer, *J. Electrochem. Soc.*, *131*:221C (1984).

298.   C. P. Andrieux and M.-M. Saveant, *J. Electroanal. Chem.*, *93*:163 (1978).

299.   F. C. Anson, *J. Phys. Chem.*, *84*:3336 (1980).

300.   R. W. Murray, *Phil. Trans. R. Soc. Lond. A.*, *302*:253 (1981).

301.   C. P. Andrieux, J. M. Dumas-Bouchiat, and J.-M. Saveant, *J. Electroanal. Chem.*, *131*:1 (1982).

302.   A. Bettelheim, B. A. White, and R. W. Murray, *J. Electroanal. Chem.*, *217*:271 (1987).

303a.   C. R. Cabrera and H. D. Abruña, *J. Electroanal. Chem.*, *209*:101 (1986).

303b.   A. I. Breikss and H. D. Abruña, *J. Electroanal. Chem.*, *201*:347 (1986).

304a.   B. P. Sullivan and T. J. Meyer, *J. Chem. Soc. Chem. Commun.*, 1984, p. 1244.

304b.   B. P. Sullivan, C. M. Bolinger, D. Conrad, W. J. Vining, R. W. Murray, and T. J. Meyer, *J. Chem. Soc. Chem. Commun.*, 1985, p. 1416.

304c.   T. R. O'Toole, L. D. Margerum, T. D. Westmoreland, W. J. Vining, R. W. Murray, and T. J. Meyer, *J. Chem. Soc. Chem. Commun.*, 1985, p. 1416.

305.   S. Cosnier, A. Deronzier, and J.-C. Moutet, *J. Electroanal. Chem.*, *207*:315 (1986).

306a.   J. Hawecker, J.-M. Lehn and R. Ziessel, *J. Chem. Soc. Chem. Commun.*, 1983, p. 536.

306b.   J. Hawecker, J.-M. Lehn a d R. Ziessel, *J. Chem. Soc. Chem. Commun.*, 1984, p. 328.

306c.   J. Hawecker, J.-M. Lehn, and R. Ziessel, *J. Chem. Soc. Chem. Commun.*, 1985, p. 56.

307.   A. R. Guadalupe, D. A. Usifer, K. T. Potts, H. C. Hurrell, A.-E. Mogstad and H. D. Abruña, *J. Chem. Soc.* (in press).

308.   H. S. White, H. D. Abruña, and A. J. Bard, *J. Electrochem. Soc.*, *129*:265 (1982).

309.   R. Noufi, *J. Electrochem. Soc.*, *130*:2126 (1983).

310.   D. L. Dubois and J. A. Turner, *J. Am. Chem. Soc.*, *104*:4989 (1982).

311.   H. Yoneyama, J. Shiota, and H. Tamura, *J. Electroanal. Chem.*, *159*:361 (1983).

312.   R. E. Malpas, F. R. Mayers, and A. G. Osborne, *J. Electroanal. Chem.*, *153*:92 (1983).

313.   R. E. Malpas and B. Rushby, *J. Electroanal. Chem.*, *157*:382 (1983).

314.   D. L. Dubois, *Inorg. Chem.*, *23*:2047 (1984).

315.   K. Rajeshwar, M. Kaneko, A. Yamada, and R. N. Noufi, *J. Phys. Chem.*, *89*:806 (1985).

316.   K. Honda and A. J. Frank, *J. Phys. Chem.*, *88*:5577 (1984).

317.   A. R. Guadalupe and H. D. Abruña, *Anal. Chem.*, *57*:142 (1985).

318.   L. M. Wier, A. R. Guadalupe, and H. D. Abruña, *Anal. Chem.*, *57*:2009 (1985).

319.   A. R. Guadalupe, L. M. Wier, and H. D. Abruña, *American Laboratory*, *18*:102 (1986).

320.   M.-C. Pham, G. Tourillon, P.-C. Lacaze and J. E. Dubois, *J. Electroanal. Chem.*, *111*:385 (1980).

321.   J. A. Cox and P. J. Kulesza, *J. Electroanal. Chem.*, *159*: 337 (1983).

322.   K. H. Lubert, M. Schnurrbush, and M. Thoman, *Anal. Chim. Acta*, *144*:123 (1982).

323.   M. J. Gehron and A. Brajter-Toth, *Anal. Chem.*, *58*:1488 (1986).

324.   R. W. Murray, A. G. Ewing, and R. A. Durst, *Anal. Chem.*, *59*:379A (1987).

325.   G. P. Kittlesen, H. S. White, and M. S. Wrighton, *J. Am. Chem. Soc.*, *106*:7389 (1984).

326.   G. P. Kittlesen, H. S. White, and M. S. Wrighton, *J. Am. Chem. Soc.*, *107*:7373 (1985).

327.   E. W. Paul, A. J. Ricco, and M. S. Wrighton, *J. Phys. Chem.*, *89*:1441 (1985).

328.   J. W. Thackeray, H. S. White, and M. S. Wrighton, *J. Phys. Chem.*, *89*:5133 (1985),

329.   M. J. Natan, T. E. Mallouk, and M. S. Wrighton, *J. Phys. Chem.*, *91*:648 (1987).

330.   M. S. Wrighton, *Science*, *231*:32 (1986).

4

Polyvinylidene Fluoride for Piezoelectric and Pyroelectric Applications

Michael A. Marcus / Eastman Kodak Company, Rochester, New York

I. INTRODUCTION

In 1969, Kawai [1] discovered that poly(vinylidene fluoride) (PVF_2) exhibits large piezoelectric and pyroelectric effects when appropriately polarized. This allows a variety of new applications to be considered which were difficult to achieve with conventional piezoelectric and pyroelectric materials. Optimization of each of these diverse applications requires differing material properties, thus spurring a large materials development effort over the last decade. The physical properties of PVF_2 are found to vary strongly according to the method of manufacture. Degree of orientation, temperature of processing, and film casting methods all affect the crystalline structure of the resultant films. Polarization methods and conditions also affect the crystallinity of the films and the resultant piezoelectric and pyroelectric coefficients, as well as the activity distributions. Methods to manufacture PVF_2 films are described, and applications are reviewed with a discussion of the desired film properties for each application.

Of all the polymers studied to date, PVF_2 and its copolymers [2-5] exhibit the largest piezoelectric and pyroelectric coefficients. Today PVF_2 is being used in a variety of transducer applications [6-10], and there are over 10 companies offering polarized PVF_2 materials for sale. In the last decade, there have been over 1000 papers published on PVF_2 and over 300 patents associated with its use [2].

PVF_2 films have been prepared by a variety of manufacturing techniques, including chill wheel extrusion, thermal lamination, and solvent casting. Films have been oriented by free-span oven stretching, roll drafting, compression rolling, and tentering. The relative amounts of each of the crystalline phases of PVF_2 is found to be strongly dependent upon the manufacturing conditions [11-13]. Analysis of the resulting films have been performed by differential scanning calorimetry (DSC), x-ray, Fourier-transform infrared spectroscopy (FTIR), dielectric constant and loss, birefrigence measurements, Instron studies, and poling with subsequent measurement of activity. Trends in those data are described, and implications for utilization of the films in transducers is discussed.

II. PROPERTIES OF PVF_2

Poly(vinylidene fluoride), PVF_2, is a semicrystalline polymer of repeat unit CH_2—CF_2. It can crystallize into at least four distinct forms, referred to as the alpha, beta, gamma, and delta phases, or Forms II, I, III, and IV respectively [2,13,14]. The alpha phase (Form II), produced by cooling from the melt, has a slightly distorted TGTG' chain conformation and is nonpolar due to antiparallel packing of adjacent chains [15]. The beta phase (Form I) is usually obtained by mechanical deformation of melt-crystallized alpha phase films, either by stretching or rolling. Form I has an extended all-trans planar zigzag conformation with an orthohombic unit cell. It has a large net dipole moment (2.1 D) and is the most important polymorph for piezoelectric and pyroelectric applications. The gamma phase (Form III) is produced by solvent crystallization [2,16,17]. It has a conformation intermediate between the alpha and beta phases, being T_3GT_3G'. The delta phase (Form IV) is formed by poling the alpha phase with fields in the order of 0.6–1.5 MV/cm [18,19]. This phase is also called the polar alpha phase since it is formed by a rotation of dipole moments normal to the c direction of alternate chains in alpha-phase PVF_2. When poled at still higher fields, the polar alpha phase is converted into the beta phase [19–21].

To produce useful transducer film, the PVF_2 film must be oriented and polarized. Unoriented PVF_2 films do not show strong piezoelectric and pyroelectric behavior. Typically, the larger the degree of orientation, the higher the resultant activities.

Recently there has been much interest in copolymers of PVF_2 [13]. Copolymers of vinylidene fluoride and trifluorethylene and tetrafluorethylene containing 50–95 mole percent VF_2 are of interest since they crystallize from the melt directly into the beta phase. This occurs because the presence of the extra bulking fluorine atoms (compared to hydrogens) does not allow the occurrence of the TGTG' conformation. The trifluorethylene-containing copolymers also exhibit a ferroelectric curie point in their dielectric properties. These polymers can be polarized well without further orientation.

III. FILM CASTING METHODS

A. Chill Wheel Extrusion

In this technique, PVF_2 resin is melted in a barrel equipped with a metering screw which extrudes molten film through a thin die slot [22]. The molten polymer is contacted by a rotating chill wheel (a smooth stainless steel roller), held at an appropriate temperature below the melting point of the polymer. The molten polymer orients as it solidifies against the chilled roller. The factors which control the film thickness and orientation are extruder screw RPM, thickness of the die opening, chill wheel RPM, and amount of necking out of the die slot. The amount of necking can be controlled by melt temperature, die-slot-to-chill-wheel contact distance, chill wheel temperature, and by the use of air pinning, electrostatic pinning, or vacuum hold-down. After the chill wheel, the film is kept under tension and wound into rolls with takeup rollers. Before final windup, the thickened edges of the films are slit so that rolls can be wound without wrinkles. Highly uniaxially oriented alpha-phase PVF_2 films have been prepared by this technique, as indicated by x-ray and FTIR. Figure 1 shows dielectric data at 1 kHz and room temperature for Solef 1008 PVF_2 films prepared by this technique with an 8" diameter chill wheel rotating at different speeds. The relative dielectric permittivity of PVF_2 films is found to increase upon orientation from 9.3 in unoriented films to 16 in highly uniaxially oriented films. The dissipation factor at 1 kHz is also found to decrease as a function of orientation.

B. Solvent Casting

PVF_2 films can be cast from solvents such as DMF (dimethylformamide), DMAc (dimethyl acetamide) and HMPA (hexamethylphosphoric triamide). Unpolarized PVF_2 can exist in at least three polymorphic forms, all three of which can be formed by solvent casting depending upon the solvent system, substrate, surfactants used (if any), and drying conditions [16]. Gamma-phase films have been prepared by casting

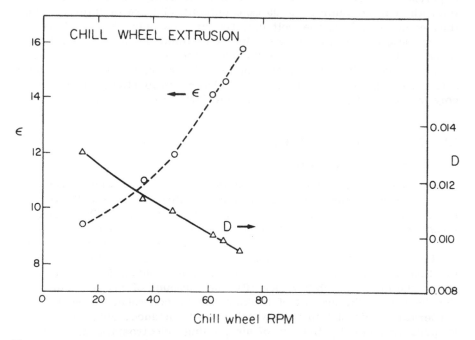

Fig. 1   Dielectric constant and dissipation factor at 1 kHz for chill wheel extruded Solef 1008 PVF_2 as a function of chill wheel RPM.

a 10% PVF$_2$ solution by weight in DMF. Beta-phase PVF$_2$ films have been prepared by precipitating a 5% by weight PVF$_2$ DMAc solution into an aqueous acetic acid bath. Figure 2 (top trace) shows FTIR data for solvent-cast Kynar 721 resin (10% by weight in DMF, dried at 50°C). This film has a dielectric constant of 9.0 at 1 kHz and a dissipation factor of 0.030 at 1 kHz. FTIR data indicate that this film is predominantly alpha-phase.

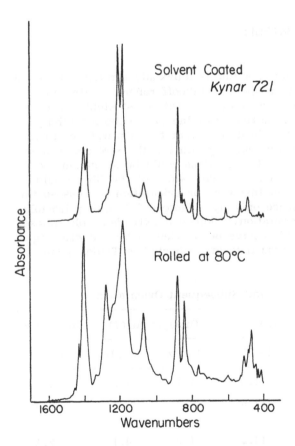

Fig. 2   Infrared absorption spectrum for solvent-cast KYNAR 721 PVF$_2$ resin cast from a 10% solution by weight in DMF (top trace). Also shown is the infrared absorption spectrum after compression rolling of this film (bottom trace).

C. Thermal Lamination

In thermal lamination, the web of interest is melt-extruded onto an already cast web. This method is useful for preparing thin films since problems associated with handling thin unsupported webs are avoided [22]. Usually the material is melt-cast from a die through a pair of nip rolls with the point of contact with the cast web being just before the nip. Alpha phase films of about 12–15 microns in thickness have been cast onto unoriented polyester (Estar) webs. The dielectric constant at 1 kHz ranges from 11–13.5 and the dissipation factor ranges from 0.012–0.016 for films produced by this technique.

IV. ORIENTATION TECHNIQUES

A. Roll Drafting

In the roll drafting technique, a film is uniaxially stretched by over-driving takeup rollers with respect to takeoff rollers. Infrared heater banks can also be used before the sets of stretching rollers to help preheat the film to improve its adhesion to the stretching rollers. The rollers are also heated so that the film will wet to the roller surface, thus minimizing slippage during the stretching operation. With available equipment PVF_2 films could be stretched by a factor of about 4.5 in length [22]. The width of the films would also decrease significantly in this type or orientation process so that film thickness would decrease only by a factor of 3 during 4.5-fold stretching for uniaxial orientation. Melt-cast extruded alpha phase and solvent coated PVF_2 films have been oriented by roll drafting at 160°C; the resulting dielectric data are reported in Tables 1 and 2.

Table 1   Melt Cast Extrusion and Subsequent Orientation

Film type	Epsilon		Dissipation (%)		Breakdown field MV/cm
	1 kHz	10 kHz	1 kHz	10 kHz	
Cast	11.8	11.5	1.7	1.9	1.8
Roll drafted (3.5X)	13.2	12.9	1.3	2.1	2.3
Rolled (3.5X)	14.6	14.2	1.9	4.1	2.1
Oven stretched (4.0X)	15.1	14.7	1.5	4.3	2.2

Table 2   Solvent Cast and Subsequent Orientation

| Film type | Epsilon | | Dissipation % | | Breakdown field |
	1 kHz	10 kHz	1 kHz	10 kHz	MV/cm
Cast	8.9	8.5	2.9	2.5	1.4
Roll drafted (4.5X)	15.4	15.1	2.1	2.5	2.2
Oven stretched (4.5X)	15.4	14.9	1.8	3.7	2.4
Rolled (4.0X)	15.5	15.1	2.0	3.6	2.3

The crystalline morphology of these oriented films was analyzed by x-ray and FTIR spectroscopy and found to be predominantly alpha phase, with a small amount of beta phase present. As seen in Tables 1 and 2, the roll-drafted melt-cast film had a lower dielectric constant and dissipation factor than the roll-drafted solvent-cast film. It is also true that the solvent-coated film had a higher degree of uniaxial orientation imparted to it than did the melt-extruded film. This would tend to increase the dielectric constant, as shown above in the rapid chill wheel extrusion case.

Another kind of drafting process called bead drafting can be performed by gripping the side edges of the film to be oriented and pulling the film in the machine direction while the film is heated. In this type of drafting process, the width of the film is forced to remain constant. PVF_2 films can also be oriented by stretching by a factor of up to 5X with the bead drafting process.

B. Oven Stretching

Uniaxial oven stretching was performed by gripping films at their ends in an oven and elongating the films. This is the typical stretching experiment performed in an Instron tensile testing apparatus. Samples were oriented by this technique at a variety of temperatures. Stretching at 80°C resulted in almost 100% conversion from the alpha to beta crystalline phase when films were stretched in excess of 3.5X their initial length. In all cases, the dielectric constant was found to increase as the degree of orientation increased. This agrees with theoretical models for beta-phase PVF_2 [23]. However, unlike for rapid chill wheel extruded films, the dissipation factor was higher in the oriented films than in their nonoriented counterparts at 10 kHz. The temperature and frequency dependence of the dissipation factor also was drastically altered in all cases by orientation.

C. Compression Rolling

In the compression rolling method, a polymer film is passed through rotating heated rollers under extremely high pressures [24]. This results in uniaxially oriented films without any change in film width. The results reported here are for compression rolling at 80°C under a force of 250,000 lb (see Tables 1 and 2 and Figs. 3–5). A 0.1% aerosol OT/water solution [24] has been applied for the nip for lubrication to increase the orientation imparted to the films during rolling. Films could be oriented up to 5X in a single pass through the rollers depending on the temperature, rolling pressure, and rolling speed. When passed through compression rollers, films increase in clarity. A crystalline phase transition also occurs as indicated by x-ray and FTIR analysis. Figure 2 (bottom trace) shows FTIR data for a rolled solvent-cast film, and Fig. 5 shows

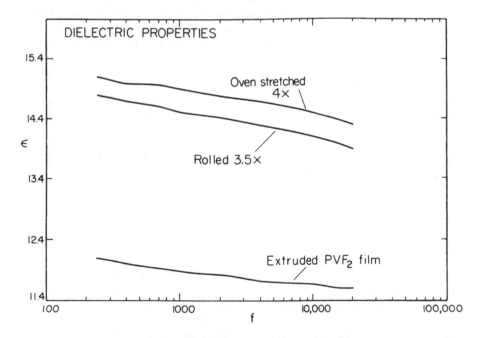

Fig. 3   Dielectric constant as a function of frequency for melt-cast extruded, 4X oven-stretched, and 3.5X rolled PVF_2 films. The latter two were oriented at 80°C.

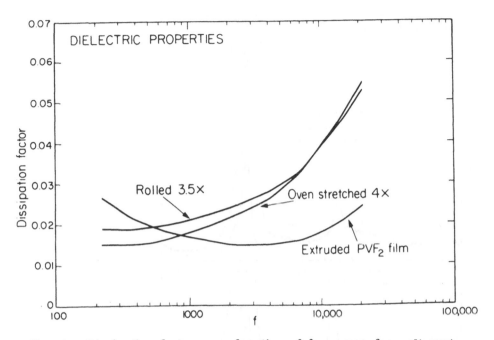

Fig. 4   Dissipation factor as a function of frequency for melt-cast extruded, 4X oven-stretched, and 3.5X rolled PVF$_2$ films. The latter two were oriented at 80°C.

x-ray data for melt-cast and rolled melt-cast films. The new crystalline structure resembles the beta phase, but the lattice spacings are slightly different (see Fig. 5). There may be a distorted lattice constant due to the high pressures involved in the compression rolling technique. Poling the rolled films alters the lattice constants to make them more like the usual beta-phase films (Fig. 5, bottom trace). As with the other orientation techniques, the dielectric constant of the rolled films increased with the amount of compression rolling (Fig. 3).

D. Tentering

Tentering is a method used to orient the films in a direction perpendicular to the machine direction. It is also called transverse direction orientation. In tentering, the film is oriented by gripping the film at its edges and stretching its width while the film is passing through an oven. A typical tentering oven has a preheat section, one or more stretching sections, and an annealing zone. Tentering typically increases the beta-phase content of PVF$_2$ films, with higher

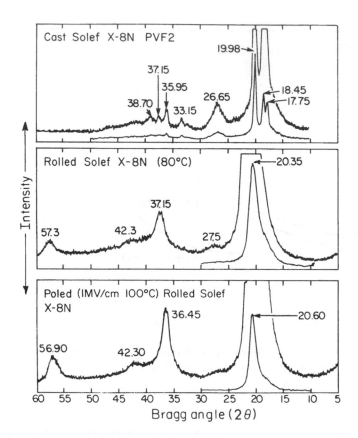

Fig. 5   X-ray absorption *versus* Bragg angle for melt-cast film, and rolled and poled film.

degrees of stretch and lower stretch temperatures favoring increased beta phase conversion of films with a higher initial alpha phase content.

V. POLING

For PVF_2 to be useful as a transducer material, the polymer must be oriented and polarized.   Unoriented PVF_2 films do not usually exhibit large piezoelectric and pyroelectric coefficients.   Typically, the larger the degree of orientation in a PVF_2 film, the higher will be the resultant activities.   The conventional thermal poling technique consists of electroding both surfaces of the polymer film, subjecting the film to electric fields of 1 MV/cm at high temperature (typically 80–100°C), followed by cooling in the presence of the applied field [25,26]. Other methods of poling including corona [21], plasma [27], and higher field room temperature poling [28].

Thermal poling at 100°C and corona poling experiments have been performed on films prepared by all of the above-mentioned techniques. It is found that d_{31} increases as the degree of orientation increases. Any of the utilized orientation techniques yields films with d_{31} values of about 25 pC/N for 4.5X uniaxially oriented films when poled to saturation. There are no noticeable differences in the results obtained for films thermally poled or corona-poled to saturation.

Structural phase transitions are observed during poling and can be seen by x-ray, FTIR, and DSC. Figure 6 shows an example of this using DSC. Here a melting curve of nonoriented melt-cast alpha-phase film is shown before and after thermal poling at 100°C for one hour

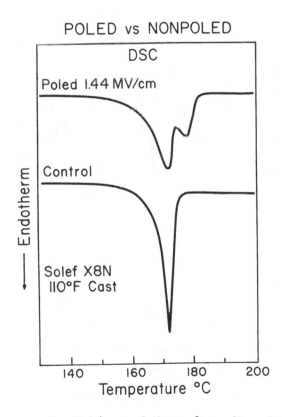

6   Melting endotherm for melt-cast and poled melt-cast film.

with a field of 1.44 MV/cm. The top trace shows the conversion to
the polar alpha phase. The poled film has a d_{31} = 4.0 pC/N. In
Fig. 7 we see FTIR spectra as a function of poling field for 9-micron
biaxially oriented films prepared by drafting and tentering. These
films were poled in an oil bath at 100°C for 20 minutes and then
quickly quenched to room temperature with the field still applied.
For a poling field of 0.46 MV/cm, g_{31} = 0.02 Vm/N; for a poling
field of 1.23 MV/cm g_{31} = 0.06 Vm/N; for 2.62 MV/cm, g_{31} = 0.09
Vm/N for these films.

The data in Fig. 7 indicate that structural phase transitions occur
during the poling of PVF_2 films. The conversion from the alpha
phase to the polar alpha phase and then to the beta phase is observed as the poling field is increased. Above a threshold of about
0.6 MV/cm, decreases in intensity associated with the alpha-phase

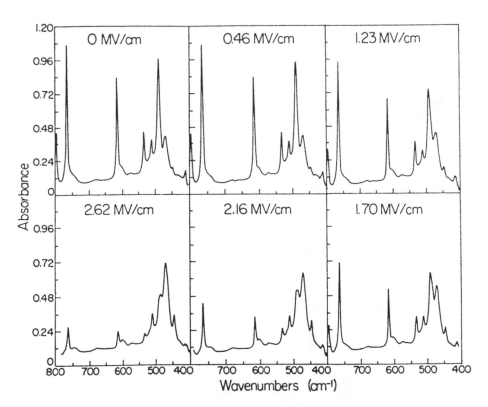

Fig. 7   FTIR absorption spectra of 9 μm biaxially oriented PVF_2
films for various poling fields.

vibrations occurred (i.e., 855 cm^{-1}, 765 cm^{-1}, 612 cm^{-1}, and 530 cm-1). Also, most bands associated with the beta phase required fields in excess of 1.5 MV/cm before their intensities are observed to increase.

In general, three classes of IR bands are evident as a function of poling field for the biaxially oriented films which were initially a mixture of alpha and beta phase PVF$_2$. These are:

1. Class A bands which decrease in intensity as the poling field is increased above a threshold value of about 0.6 MV/cm
2. Class B bands which decrease slightly in intensity at a similar threshold poling field, but start increasing in intensity again as the field increases above 1.5 MV/cm.
3. Class C bands which increase in intensity above a threshold value of 0.6 MV/cm.

The class A modes can all be assigned to the alpha phase, whereas the class B and C modes are assigned to the beta phase. Transition moment arguments for different types of vibrations can be used to explain the differences in behavior between class B and class C bands.

VI. TRANSDUCER APPLICATIONS

PVF$_2$ offers several advantages over more traditional ceramic and crystalline transducer materials. It is easy to produce in large sizes and can readily be machined into complex shapes. It can easily be cut without damage to the transducer, and it can be subjected to large mechanical shocks and vibration amplitudes. It is lightweight, flexible, has a large bandwidth and low Q, and is an excellent acoustic match to water and biological tissue. In many applications, complex transducer mounts are eliminated by simply gluing the transducer to the surface under investigation.

PVF$_2$ also has some disadvantages relative to the more tranditional transducer materials. It has low piezoelectric strain coefficients and low electromechanical coupling factors. This is shown in Table 3 where we compare many transducer properties of PVF$_2$ to those of some other common piezoelectric and pyroelectric materials. Although commercially poled films as thick as 1 mm are available, it is difficult to pole thick films. If a higher voltage output is required, several sheets may be stacked. PVF$_2$ is also subject to thermal depolarization when heated above 100°C, thus limiting its use at these temperatures.

When considering PVF$_2$ for use in a particular transducer application, the designer should remember that film thickness, orientation, and crystalline structure affect its suitability for a given application.

Table 3   Comparison of Properties of Some Piezoelectric/Pyroelectric Materials

Materials	d_{31}	d_{32} (pC/N)	d_{33}	g_{31}	g_{33}	p ($\mu C/m^{-2}K$)	Epsilon
PVF$_2$ (uniaxial)	20	3	-35	.18	$-.31$	-27	15
PVF$_2$ (biaxial)	9.8	9.8	-39	.091	$-.36$	-40	12
BaTiO$_3$	-78	-78	190	$-.005$	.012	200	1260
BZT-5	-171	-171	374	$-.01$	.024	300	1700
TGS	—	—	—	—	—	350	43

Table 4   Desirable Film Type for Various Applications

Application	Film type
Flexure-Mode Devices	Thin, uniaxially stretched
Displacement Transducers	Thick, uniaxially stretched
Hydrophones	Thick, biaxially stretched
Pyroelectric Detectors	Thin, biaxially stretched
Capacitors	Thin, high dielectric constant

Table 4 lists a variety of applications along with the desirable film characteristics. For flexure mode devices such as vibrational fans [29], in which an electrical input is converted into mechanical motion of a cantilever beam, thin, uniaxially oriented film is desired. The induced deflection of the tip of a cantilever beam for a given voltage input is proportional to d_{31} and inversely proportional to the square of the thickness of the piezoelectric film [30]. For a displacement the induced voltage is proportional to d_{31} and the square of the thickness [6]. Thicker material also allows operation at higher frequencies for transducers of the same length. Thus, thick uniaxially oriented transducer film is desirable in these types of applications. In pyroelectric applications we wish to have a large pyroelectric coefficient and a short transit time through the film. In most cases, biaxially oriented film has a higher pyroelectric coefficient than uniaxially oriented films, implying that thin biaxially stretched films are desirable for pyroelectric applications. For hydrophones based on a hydrostatic principle the induced voltage is proportional to the hydrostatic piezoelectric coefficient d_h and the thickness of the film. This implies that thick biaxially oriented films are desirable for these applications. In using an actual transducer film for this kind of application, the user must be cautioned that even though a piezoelectric film is quoted as having high piezoelectric coefficients, it may end up having a very low hydrostatic coefficient depending upon the actual film orientation. This is because $d_h = d_{31} + d_{32} + d_{33}$, and d_{33} has the opposite sign of d_{31} and d_{32}. For capacitor applications, thin films with high dielectric constant and high breakdown strength are desirable.

Another often-overlooked feature in determining PVF_2 transducer device performance is the uniformity of polarization. If the polarization is not uniform in the thickness dimension of the film, then device performance depends upon the way in which the film is used, that is, upon which side of the film is grounded and/or clamped.

If an asymmetrically poled film is used as a pyroelectric detector, the high-activity side will respond faster to a change in the temperature than the low activity side. If a series bimorph flexure mode device is constructed, a greater amount of bending will occur when the high-activity sides of the film face outward as compared to when they face inward. It is possible to control the piezoelectric activity distribution in PVF_2 films [31], and novel devices can be constructed using asymmetrically poled film.

Applications of PVF_2 are now discussed by grouping them according to transduction mechanism. Tables 5, 6, and 7 list a variety of pyroelectric, electro-mechanical, and mechano-electric applications for PVF_2. Table 8 lists some commercialized transducer applications of PVF_2.

A. Pyroelectric Transducers

Table 5 describes a number of pyroelectric applications for PVF_2. Pyroelectric detection involves the absorption of incident infrared radiation by the detector material. When heat is absorbed, a pyroelectric signal is induced which is proportional to temperature change rather than absolute temperature. Optimization of pyroelectric

Table 5 Pyroelectric Applications of PVF_2

Infrared Detectors
Vidicon Targets
Thermal Imaging Arrays
Automatic Room Light Controllers
Laser Beam Profiling
Radiometers
Heat Scanners
Intrusion Detectors
Spectral Reference Detectors
Reflectometers
Photocopying
Charge Separator, Filter
Heavy Nucleii Detectors
Micro-Calorimeters

Table 6   Electro-Mechanical Applications of PVF_2

Loudspeakers
  Ballons, caps, cylinders, cones, headphones

Flexure Mode Transducers
  Deformable mirrors
  Electronic displays
  Light deflectors
  Optical scanners
  Position sensors
  Variable aperature diaphrams
  Venetian blinds
  Vibrational fans

Modulators for Fiber Optics

Surface Acoustic Wave Devices

Adaptive Optics

Ultrasonic Transmitters

Optical Phase Modulation

Antifouling Vibration Drivers

Gas Flow Meter

Acoustic Microscope

Heat Exchange Membrane

Vibrating Membranes

Tone Generators

---

detector performance depends upon materials parameters. Desirable properties for a pyroelectric detector include (a) large pyroelectric coefficient, (b) high volume resistivity, (c) low dielectric constant and dissipation factor, (d) low specific heat, (e) low density, and (f) a broad usable temperature range. PVF_2 has a relatively low materials figure of merit in comparison to some of the other more common pyroelectric materials. However, it is being used today because of its flexibility, chemical stability, mechanical strength, power-handling ability, and ease of preparation of extremely thin films. Today PVF_2 is being used in a variety of commercialized devices which make use of the pyroelectric effect. It is used in a number of burglar alarms as a passive infrared detector [32]. A room light controller called "Infracon" has been manufactured by United

Table 7   Mechano-Electric Applications of PVF_2

Microphones	Fuses
Noise Cancelling	Impact Detectors
Phonograph Cartridge Pickups	Coil Sensors
Hydrophones	Traffic Sensors
Sound-field mapping	Personal Verification Device
Naval	Blood Flow Monitor
Imaging	Dust Particle Detectors
Arrays	Wind Driven Power Generators
Medical Ultrasonic Imaging	Wave Driven Power Generators
Pulse Monitors	Tactile Sensors for Robotics
Touch Switches	Munition Detonators
Accelerometers	Physiological Implants
Stress and Strain Gauges	Pacemakers
Shock and Vibration Gauges	
Nip Pressure Transducers	

Table 8   Commercialized Applications

Intrusion Detectors

Loudspeakers

Microphones

Heat Scanners

Room Light Controllers

Gas Flow Meters

Infrared Detectors

Headphones

Hydrophones
  Scanning sonar
  Field mapping

Technologies and Tishman Research Corporation. New England Research Center markets a PVF_2 pyroelectric detector with detectivity on the order of $D^* = 1 \times 10^8$ cmHz$^{1/2}$/W at 10 Hz and 10.6 microns incident wavelength. PVF_2 has also been used as a vidicon tube target material [33], as a laser beam profiler [34], as the active medium for pyroelectric copying [33], and as a cosmic dust detector [36,37]. In fact, PVF_2 detectors are on board the USSR Vega spacecraft launched December 15 and 21, 1984, for encounters with Halley's comet [37].

B. Electro-Mechanical Transducers

Table 6 lists a variety of electro-mechanical transducer applications for PVF_2. Electro-mechanical transducers utilize either longitudinal (thickness mode) or transverse piezoelectric effects. Ultrasonic transducers usually utilize the longitudinal piezoelectric effect, whereas flexure-mode devices and airborne sound transducers utilize the transverse piezoelectric effect. Loudspeakers [38,39] and headphones [38] were among the first commercialized applications for PVF_2. Pioneer has marketed PVF_2 tweeters with a deviation in frequency of less than ±2 dB over the frequency range of 2–20 kHz [38]. Many applications have been developed based on the theory of flexure in beams [6,40] and some novel structures have been described by Linvill [41]. Flexure mode devices are usually constructed in a bimorph or multimorph construction. These devices are constructured so that when an electric field is applied to the device, the part above the neutral axis will tend to deform in one direction, whereas the part below the neutral axis will tend to deform in the opposite direction. This results in large deflections of the free end of a cantilever-mounted structure. Tip deflection of a cantilever beam is proportional to d_{31}, the applied voltage, the square of the length of the cantilever, and the inverse square of the thickness. The resonance frequency is proportional to thickness and the inverse square of the length. If we construct multimorphs with 2N layers, then displacement and force are given by the corresponding bimorph values multiplied by 1/N and N^2 respectively. A materials figure of merit called the deflection-bandwidth product (DBWP) [30] is useful when considering applications requiring both large deflections and high-speed operation. PVF_2 has a DBWP = 0.3 mHz, which is about a factor of 2 higher than any other presently available piezoelectric material. A potential large-scale application utilizing electromechanical conversion in PVF_2 is in the area of marine antifouling [42]. Other electromechanical applications for PVF_2 include fans [29], displays [43], telephone tone ringers [44], and ultrasonic generators [45–46]. Medical ultrasonic transducers have been constructed using PVF_2 and copolymers which have a broader bandwidth than PZT transducer counterparts [47].

C. Mechano-Electric Transducers

Table 7 lists a number of mechano-electric transducer applications for PVF_2. Microphones and phonograph cartridges [38] were among the first applications for PVF_2 transducer films. Various workers have examined the use of PVF_2 for hydrophones [34] and medical imaging applications [7,48]. Today, Marconi [9] markets membrane hydrophones for ultrasonic field measurements as well as a 100-element 360° scanning sonar transducer for use in their underwater 360° Surveillance Sonar System. Gas flow meters are being manufactured by Perkin Elmer in which a PVF_2 band transmits an acoustic wave down a tube to a PVF_2 band receiver. The receiver and transmitter are then interchanged, and the difference in time of flight can be used to calculate the gas flow velocity. A few potential large-scale applications for PVF_2 include wind power generation [49] and water wave power generators [50]. Many applications for PVF_2 sensors are found in the medical field. PVF_2 has been used in devices which provide electrical impulses in an implant to promote bone healing [51]. Miniature piezoelectric polymer hydrophones have been developed [52] which are useful as a medical imaging reference and for in vivo ultrasonic measurements of tissue properties. PVF_2 catheter-tip transducers have been developed to measure intra-cavitary pressure and sound [53], and these devices are found to have much better transient response than silicone strain gauges. Insole pressure transducer arrays have been found useful in detecting orthopedic and neurological defects in patients' feet. PVF_2 sensors for monitoring the behavior of prosthetic vascular grafts [55] as well as fetal heart monitors [56] have been developed. Experiments have been performed to develop a sensorized skin-like tactile transducer for prostheses [57]. The experiments indicate that it may be feasible to use ferroelectric polymers as the transducer material.

D. Micellaneous Applications

Optical-to-mechanical transducers have been developed utilizing PVF_2. In one such device a photoconductor with a top transparent electrode is bonded to a PVF_2 sheet to produce a unimorph type structure. A bias voltage is applied to the photoconductor, and, when illuminated, the field appears across the PVF_2 layer, causing it to bend. This device can be used as a light-controlled variable aperture or an automatic exposure mechanism in a camera. Second harmonic generation has also been demonstrated in PVF_2 [3]. Thermomechanical devices and mechanical transformers are also conceivable utilizing PVF_2.

Today PVF_2 is finding widespread use in many diverse disciplines, including acoustics, ultrasonics, pyroelectric detection, biomedical engineering, and nondestructive testing. New transducer designs are still needed, and new applications are being sought which require large amounts of transducer film. Once the designer has a clearer picture of the unique properties of PVF$_2$ as compared to more conventional transducer materials, the new transducer applications will be developed.

VII. SUMMARY

A variety of methods of manufacturing oriented and nonoriented PVF$_2$ films have been described. These include chill wheel extrusion, solvent casting, thermal lamination, drafting and tentering, oven stretching, and compression rolling. In all cases, the resultant film properties are strongly dependent upon the degree of orientation and manufacturing processing conditions. In general, increasing the degree of orientation increases the dielectric constant and piezoelectric and pyroelectric properties of polarized films. A given orientation-increasing technique may yield similar improvements in these properties even when applied to starting materials prepared by different methods.

Applications for PVF$_2$ films have also been reviewed, and it is was shown that different film properties are desirable for different applications.

REFERENCES

1. H. Kawai, *Jpn. J. Appl. Phys.*, 8:975 (1969).
2. A. J. Lovinger, *Developments in Crystalline Polymers*, Vol. 1, D. C. Bassett, ed., Applied Science, London, 1982, p. 195.
3. T. Furukawa, G. E. Johnson, H. E. Bair, Y. Tajitsu, A. Chiba, and E. Fukada, *Ferroelectrics*, 32:61 (1981).
4. Y. Higashihata, J. Sako, and T. Yagi, *Ferroelectrics*, 32:85 (1981).
5. T. Furukawa, M. Date, M. Ohuchi, and A. Chiba, *J. Appl. Phys.*, 56:1481 (1984).
6. M. A. Marcus, *Ferroelectrics*, 40:29 (1982).
7. H. Sussner, *Ultrasonic Symposium Proceedings*, 9:491 (1979).
8. G. M. Sessler and J. E. West, *Topics in Applied Physics— ELECTRETS*, G. M. Sessler, ed., Springer Verlag, New York, 1980, p. 347.
9. H. R. Gallantree, *Marconi Review*, 45:49 (1982).
10. G. M. Sessler, *J. Acoust. Soc. Am.*, 70:1596 (1981).
11. R. J. Shuford, A. F. Wilde, J. J. Ricca, and G. R. Thomas, *Polym. Eng. Sci.*, 16:25 (1976).
12. P. T. A. Klasse and J. Van Turnhout, *Dielectric Materials Measurements and Applications*, Bamber Press Ltd., 1979, p. 411.

13. A. J. Lovinger, *Science*, *220*:115 (1983).
14. R. Hasegana, Y. Takahashi, Y. Chatiani and H. Tadokoro, *Poly. J.*, *3*:600 (1972).
15. M. H. Bachmann and J. B. Lando, *Macromolecules*, *14*:40 (1981).
16. W. M. Prest Jr. and D. J. Luca, *J. Appl. Phys.*, *49*:5042 (1978).
17. M. Kobayashi, K. Tashiro, and H. Tadokoro, *Macromolecules*, *8*:158 (1987).
18. D. Naegele, D. Y. Yoon, and M. G. Broadhurst, *Macromolecules*, *11*:1297 (1978).
19. G. T. Davis, J. E. McKinney, M. G. Broadhurst, and S. C. Roth, *J. Appl. Phys.*, *49*:4996 (1978).
20. J. P. Luongo, *J. Polym. Sci. A2*, *10*:1119 (1972).
21. P. D. Southgate, *Appl. Phys. Lett.*, *28*:250 (1976).
22. M. A. Marcus, *Proc. 5th Int. Symp. Electrets*, Heidelberg, 1985, p. 894.
23. M. G. Broadhurst and G. T. Davis, *Topics in Applied Physics-ELECTRETS*, G. M. Sessler, ed., Springer Verlag, New York, 1980, Chap. 5.
24. R. F. Williams, Jr. and E. D. Morrison, *SPE Journal*, *27*:42 (1971).
25. P. E. Bloomfield, R. A. Ferren, P. F. Radice, H. Stefanou, and O. Sprout, *Nav. Res. Rev.*, *31*:1 (1978).
26. N. Murayama, K. Nakumura, H. Obara, and M. Segawa, *Ultrasonics*, *14*:15 (1976).
27. J. E. McKinney, G. T. Davis, and M. G. Broadhurst, *J. Appl. Phys.*, *51*:1676 (1980).
28. J. M. Kenney and S. C. Roth, *J. Res. Nat. Bur. Stand.*, *84*:447 (1979).
29. M. Toda, S. Osaka, and E. O. Johnson, *RCA Engineer*, *25*:24 (1979).
30. J. K. Lee and M. A. Marcus, *Ferroelectrics*, *32*:93 (1979).
31. M. A. Marcus, *Ferroelectrics*, *57*:203 (1984).
32. J. Cohen, S. Edelman, and C. Verretti, *Nat. Bur. Stand. (US) Tech. News Bull.*, *56*:52 (1972).
33. Y. Hatanaka, S. Okamoto, and R. Nishida, *Adv. Electron. Electron. Phys.*, *52*:31 (1979).
34. H. R. Gallantree and R. M. Quillian, *Marconi Rev.*, Fourth Quarter, 1976, p. 189.
35. J. G. Bergman, G. R. Crane, A. H. Ballman, and H. M. O'Bryan, *Appl. Phys. Lett.*, *21*:497 (1972).
36. J. A. Simpson and A. J. Tuzzolino, *Nuclear Instruments and Methods in Phys. Res. A.*, *236*:178 (1985).
37. M. A. Perkins, J. A. Simpson and A. J. Tuzzolino, *Nuclear Instruments and Methods in Phys. Res. A.*, *239*:310 (1985).
38. M. Tamura, Y. Yamaguchi, T. Oyaba, and T. Yoshimi, *J. Audio Eng. Soc.*, *23*:21 (1975).

39. F. Micheron and C. Lemonon, *J. Acoust. Soc. Am.*, *64*:1720 (1978).

40. M. Toda, *Ferroelectrics*, *32*:127 (1981).

41. J. G. Linvill, *Stanford Electronics Laboratory*, *Technical Report 4834-3*, March 1978.

42. M. Latour, O. Gverlorget, and P. V. Murphy, *Studies in Electrical and Electronic Engineering*, Vol. 2, Elsevier, N.Y., 1979, p. 175.

43. M. Toda and S. Osaka, *Proc. Soc. In. Disp.*, *19*:35 (1978).

44. P. V. Murphy and G. C. Mauror, *Proc. 5th Int. Symp. Electrets*, Heidelberg, 1985, p. 783.

45. C. Alquie, J. Lewiner, and C. Friedman, *Appl. Phys. Lett.*, *29*:69 (1976).

46. P. T. A. Klaase, *Ferroelectrics*, *60*:215 (1984).

47. H. Ohigashi, K. Koga, M. Suzuki, T. Nakanishi, K. Kimura, and N. Hashimoto, *Ferroelectrics*, *60*:263 (1984).

48. R. G. Swartz and J. D. Plummer, *IEEE Trans. Sonics Ultrason.*, SU-27:295 (1980).

49. V. H. Schmidt, M. Klakken, H. Darejeh, *Ferroelectrics.*, *51*: 765 (1983).

50. G. W. Taylor and J. R. Burns, U.S. Patent 4,404,490, September 13, 1983.

51. J. J. Ficat, G. Escourrou, M. J. Fauran, R. Durroux, P. Picat and Lacabanne, *Ferroelectrics*, *51*:781 (1983).

52. P. A. Lewin, *Ferroelectrics*, *60*:127 (1984).

53. P. Dario, D. DeRossi, R. Bedini, R. Francesconi and M. G. Trivella, *Ferroelectrics*, *60*:149 (1984).

54. A. Pedotti, R. Assente, G. Fusi, D. DeRossi and C. Domenici, *Ferroelectrics*, *60*:163 (1984).

55. P. D. Richardson, P. M. Galleti, P. Dario, *Ferroelectrics*, *60*: 175 (1984).

56. F. Steenkeste, Y. Moschetto, M. Boniface, P. Ravinet and F. Micheron, *Ferroelectrics*, *60*:193 (1984).

57. P. Dario, D. DeRossi, C. Giannottie, F. Vivaldi, P. C. Pinotti, *Ferroelectrics*, *60*:199 (1984).

5

Excited States of Conjugated Polymers

Z. G. Soos and G. W. Hayden / Princeton University, Princeton,
New Jersey

I. INTRODUCTION

A. π-Electrons and Excitations of Conjugated Polymers

The electronic structure of atoms, molecules, and solids follows in principle from Schrödinger's equation. Generations of physical scientists have by now developed numerous approximation schemes, some refined into practically separate disciplines. The requisite accuracy depends on the issues addressed and inevitably increases with experimental, theoretical, and computational advances. Different approaches to conjugated polymers are summarized in recent reviews [1–4] and conference proceedings [5–8]. While some unification is probable, different aspects of conjugated polymers will probably continue to require separate descriptions.

Conjugated polymers may be considered, for example, to be extremely large molecules and thus susceptible to the techniques of theoretical chemistry. They may alternatively be seen as one-dimensional periodic arrays and analyzed by band-theoretical methods for metals and semiconductors. Special features of their low-lying electronic states are perhaps best understood in terms of idealized electron-phonon, nonlinear, or field-theoretical models. Phenomenological approaches based on molecular-exciton theories, or on exchange models for magnetic insulators, have also been applied. Conjugated polymers also have configurational degrees of freedom characteristic of saturated polymers. The most appropriate approach clearly depends on the problems addressed.

The discussion below is largely restricted to the low-lying electronic states of the simplest conjugated polymers. These are the states responsible for novel electrical properties that distinguish conjugated polymers from many saturated polymers. *Trans*-polyacetylene (PA) [1,5] in Fig. 1 becomes a good conductor on ∿1% doping by either electron donors or acceptors and serves as either the anode or cathode in electrochemical applications. Polydiacetylenes (PDAs) [7] in Fig. 2 are excellent photoconductors and

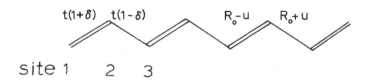

site 1   2   3

Fig. 1 Schematic representation of an all-trans polyene or *trans*-PA. Bond length alternation $R_0 \pm u$ leads to alternating transfer integrals $t(1 \pm \delta)$ in Eq. (3). All bond angles are $2\pi/3$ in the idealized structure.

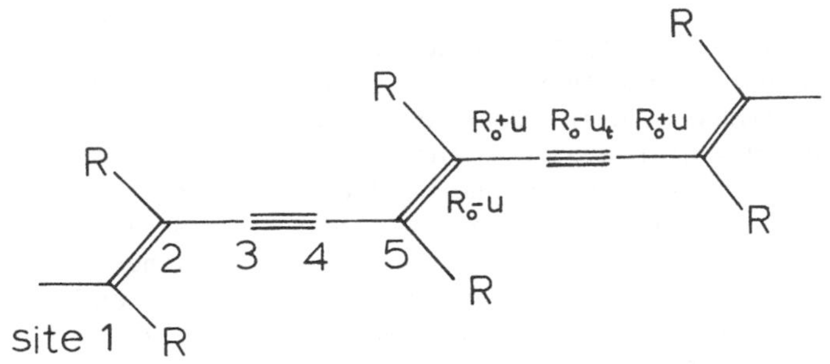

Fig. 2 Schematic representation of the polydiacetylene backbone, with nominally single, double, and triple bonds and bond angles $2\pi/3$ at sp^2 carbons, π at sp carbons. The R groups are connected by σ-bonds.

occur as single crystals for several choices of R. Both PA and PDAs show strong and potentially useful nonlinear electric suscepti-bilities [5,9]. Broadly similar electrical properties have been found in conjugated polymers based on aromatic and heterocyclic molecules [3,5]. These properties are principally associated with π-electrons in conjugated networks and may be modeled without explicitly in-cluding σ-electrons.

Conjugated molecules have been central to theoretical chemistry ever since the seminal contributions of Hückel, Coulson, Longuet-Higgins, and Pople. The development of molecular orbital methods is discussed in Salem's excellent book [10]. Related solid-state models [11] for metals and semiconductors were introduced inde-pendently by Bloch, Heisenberg, Van Vleck, and Hubbard. The resulting quantum cell models encompass many problems from physics and chemistry, can occasionally be solved exactly, and provide a consistent approximate framework for both electron-electron (e-e) and electron-phonon (e-p) contributions. A theme of this review is to present and contrast different approaches to conjugated polymers in chemistry and physics. We also discuss advances for including e-e and e-p contributions for low-energy states of PA or PDA.

The connection between π-electron systems, s-bands, localized spins, or other quantum cell models is the restriction to a few, usually one, valence orbital ϕ_p per site. In planar conjugated molecules or polymers, ϕ_p is essentially the $2p_z$ atomic orbital and is odd under reflection in the molecular plane. The exact form of ϕ_p is immaterial, since the relevant matrix elements can be fixed experimentally. Such a semiempirical approach may be supplemented by direct computations in small molecules, but are impractical in polymers, for small exchange constants in magnetic insulators, or for narrow bands of π-radical crystals in which ϕ_p is the highest-occupied molecular orbital (HOMO) of a medium-sized molecule.

There is no rigorous justification for σ-π separability or for the restriction to π-electrons. All electrons are included in current work on small molecules with large computers, and progress towards polymer applications are sketched in Section III.B. The optical gap, E_g, is about 2 ev in conjugated polymers and corresponds to a π-π^* excitation. The optical gap of small molecules is several times higher, usually in the UV. The energy mismatch between π and σ excitations provides some hope for a π-electron description of low-lying polymeric states. Charge delocalization in conjugated molecules tends to minimize local geometrical changes on excitation, at least in comparison to small molecules. The geometrical changes associated with π-electron excitation or ionization are the solitons and polarons discussed in Sections II.B and II.C.

Many different approximations are possible within π-electron theories. The most general formulation is the Pariser-Parr-Pople (PPP) model [10], which formally corresponds to an extended or modified Hubbard model [12]. The simplest and most convenient are Hückel or tight-binding models in which both electron-electron and electron-phonon interactions are neglected. It is often advantageous to combine the insights of Hückel theory with the physical realism of PPP theory. Simplicity and accuracy are effectively combined in introductory textbooks, where transparent derivations based on idealized models are routinely augmented by numerical results and experimental comparisons reflecting the most comprehensive development along the same lines. This strategy is pursued here for low-lying excited states, with special attention to novel features associated with electron-phonon and electron-electron interactions.

B. Scope of Review

The electronic structure of conjugated polymers poses new variations on many old themes. Much insight can be gained from ideas and methods in apparently unrelated fields. But conjugated polymers also present new challenges beyond creative transpositions of ideas. We begin with the unifying aspects of quantum cell models. Section I.C introduces the Hückel, Hubbard, and PPP models for π-electrons and defines their microscopic parameters. More general solid-state models are summarized in Section I.D and related to a localized representation of valence bond (VB) diagrams. Various roles of semiempirical models are discussed. Different approximations for electron-electron correlations are sketched in Section I.E. We do not review such methodologies, but merely consider their characteristic predictions.

Single polymeric strands provide the next set of topics. Such idealized models yield essentially all the available exact results for either electron-phonon or electron-electron effects. The Peierls instability of the infinite polyene is developed in Section II.A together with the Jahn-Teller instability of 4n rings, and the role of e-e interactions is summarized. Solitons in the Su-Schrieffer-Heeger (SSH) model [13] are introduced in Section II.B to illustrate how e-p coupling leads to different ground- and excited-state geometries. The advantages of the continuum limit are discussed in Section II.C for features that are large compared to bond lengths. Correlation contributions to the optical gap and to various gap states are summarized in Section II.D.

We then turn to applications, often preliminary or incomplete, involving more realistic models of polymers. Chemical differences between conjugated polymers minimally require a quantum-chemical description of the unit cell, as summarized in Section III.B, even within an idealized, single-strand picture. The parametrization of quantum cell models is related in Section III.C to molecular systems that afford richer and more precise data. The consequences of finite chain lengths, of different polymer conformations, and of interchain interactions are sketched in Section III.D, with special emphasis on necessary modifications to infinite idealized strands. A phenomenological approach to vibronic coupling due to backbone vibrations is summarized in Section III.E and related to the question of e-e contributions. Low-energy states of PDA are contrasted in Section III.F with PA excitations and analyzed within an extended quantum cell model.

Satisfactory explanations on one level may not work on another and advances in understanding invite further improvements. We will cite throughout relevant experimental results that guide choices for models or parameters. As in kinetic schemes, one observation can exclude a simple model or force extensive reassessments, but accounting for several observations may not suffice. We believe that current models work best for separate issues and have so organized the discussion. General comparisons for polymer doping, polymer morphology, conformational degrees of freedom, or interstrand interactions are premature, although such extensions of quantum cell models are in progress.

C. Hückel, Hubbard, and PPP Models

Hückel and tight-binding theories were extraordinarily useful semiempirical or phenomenological schemes long before serious derivations or *ab initio* calculations were practical. The $2p_z$ atomic orbitals ϕ_p for the pth carbon are taken to be orthonormal, a step formally justified subsequently by using Wannier functions. Other orbitals are kept in simple band theory for metals. The diagonal (Hückel α) elements of the Hamiltonian

$$e_p = <\phi_p |H| \phi_p> \simeq -I_p \tag{1}$$

are approximated in terms of an ionization potential, thus obviating the need for specifying H or doing the one-electron integral. All site energies are equal and conventionally chosen as the reference for the PA backbone in Fig. 1. Different e_p could occur for sp and sp^2 centers of PDA in Fig. 2, but variations are usually neglected both for simplicity and to avoid additional parameters.

One-electron integrals describing electron transfers or resonance between adjacent (bonded) sites are central to Hückel theory,

$$t_p(R) = <\phi_p |H| \phi_{p+1}> \tag{2}$$

Here p and p + 1 are adjacent sites separated by R along the polymer backbone. The benzene bond length $R_0 = 1.397$ A is usually chosen as a reference and $t(R_0) = -2.40$ ev is obtained from experiment [10,14]. Some 10% variations in the choice of transfer integrals is typical and unimportant as long as used consistently. The general expectation of an exponential dependence [15] in R is frequently approximated by the more convenient linear relation

$$t(R) = t + \alpha(R - R_o) \tag{3}$$

with $t = t(R_0)$ and electron-phonon coupling constant α. Finally, transfers between second and more distant neighbors are neglected, as are all two-electron terms involving interactions among π-electrons.

In second-quantized notation, with $e_p = 0$ along the polyene chain, the Hückel Hamiltonian is

$$H_t = \sum_{p\sigma} t(R_p) (a^+_{p\sigma} a_{p+1\sigma} + a^+_{p+1\sigma} a_{p\sigma}) \tag{4}$$

The fermion operators $a^+_{p\sigma}$ ($a_{p\sigma}$) create (annihilate) a π-electron with spin σ at site p. Thus H_t describes electron transfers with spin conservation between bonded sites. The total number of π-electrons is conserved,

$$N_e = \sum_p n_p = \sum_{p\sigma} a^+_{p\sigma} a_{p\sigma} \tag{5}$$

We have one π-electron per carbon in PA, which is then described as half-filled. For N sites, we obtain N × N secular determinant

whose roots are the Hückel energies λ_k, $k = 1, 2, \ldots, N$, and
whose eigenfunctions give the proper linear combinations of ϕ_p. The
same procedure holds for arbitrary $\{e_n\}$, for arbitrary topology of
bonded sites, and indeed for additional transfers between other
neighbors. Streitweiser [16] has compiled Hückel results for many
organic molecules. The PA analysis of Longuet-Higgins and Salem
[15] and subsequently of Su, Schrieffer, and Heeger (SSH) [13]
amounts to alternating transfer integrals in Eq. (4) for $N_e = N$.
Wilson [17] naturally introduced three transfer integrals for the
single, double, and triple bonds of PDA.

Hückel approximations for molecules could be tested by the 1950s.
Both the restriction to π-electrons and the neglect of electron-
electron interactions were sources of concern [10]. The quantum-
chemical route to the PPP model is summarized by Salem [10] and
goes through the zero-differential-overlap (ZDO) approximation [18],
which rationalizes the neglect of three- and four-center integrals
over the $\{\phi_p\}$. Since Hückel theory retains all one-center integrals
like Eq. (1) and (2) we are left with two-center integrals. In the
different context of d-electrons in transition metals, Hubbard [19]
argued that the repulsion of two electrons in the same ϕ_p is important even when long-range Coulomb interactions are shielded.
The on-site or Hubbard contribution is

$$H_{Hu} = \sum_p U_p n_p (n_p - 1)/2 = \sum_p U_p a^+_{p\alpha} a^+_{p\beta} a_{p\beta} a_{p\alpha} \qquad (6)$$

$$U_p = \langle \phi_p(1)\phi_p(2) | e^2/r_{12} | \phi_p(1)\phi_p(2) \rangle$$

The two electrons in ϕ_p are spin-paired. The disproportionation
reaction

$$2X \rightarrow X^+ + X^- \qquad (7)$$

identifies U to be $I - A$ for the gas-phase ionization potential and
electron affinity. These ideas were derived independently several
times over. The remaining two-electron integrals over the $\{\phi_p\}$
describe Coulomb or other spin-independent intersite interactions,

$$V = 1/2 \sum'_{pp'} V_{pp'} (n_p - z_p)(n_{p'} - z_{p'}) \qquad (8)$$

$$V(R_{pp'}) = V_{pp'} = \langle \phi_p(1)\phi_{p'}(2) | e^2/r_{12} | \phi_p(1)\phi_{p'}(2) \rangle$$

For carbons, we have $z_p = 1$ since the core charge is 1 when ϕ_p
is empty. The important aspect of Eq. (8) is the proportionality

to $n_p n_{p'}$, which is emphasized by arbitrary magnitudes and signs for the constants $V_{pp'}$ in extended Hubbard models. Within PPP theory, however, $V(R_{pp'})$ goes as e^2/R for distant sites and reduces to U_p at $R = 0$. The Ohno formula [20]

$$V(R) = U(1 + R^2/e^4 U^2)^{-1/2}$$ (9)

provides a convenient and consistent interpolation based on a single parmater U.

The PPP Hamiltonian is the sum of the Hückel contribution, H_t, in Eq. (4), the on-site repulsion H_{Hu}, in Eq. (6), and the intersite contributions, $V_{pp'}$ given in Eq. (8). Standard parameters are collected in Table 1. Their different motivation is deffered to Section III.C. Other [21] choices for U_p and some spin-independent $V_{pp'}$ occur in extended or modified Hubbard models for a wide variety of inorganic, organic and illustrative systems.

D. Quantum Cell Models: Localized Representations

We have glossed over the entirely different roles of PPP models in theoretical chemistry and of Hubbard models in solid-state physics. PPP approximations are already unfashionably restrictive for small molecules. Quantum-chemical approaches to conjugated polymers usually aim for more. Solid-state descriptions of PA excitations within the SSH model, by contrast, are fundamentally one-electron in nature. The PPP model then appears to be unnecessarily complicated. An important clue to this divergence is the dual role of

Table 1   Parameter Values for Quantum Cell Models

	$t(R_o)$ (ev)	α (ev/A)	K (ev/A^2)	U (ev)
SSH	2.5	4.1	21.0	
PPP/molecular	2.40	3.21	24.6	11.26
Vanderbilt-Mele	3.0	8.0	68.6	

The transfer integral t and electron-phonon coupling α are defined in Eq. (3), the harmonic force constant K in Eq. (17), and the Ohno potential U in Eq. (9).
Bond lengths (A): $R_o = 1.397$; $R(C-C) = R_o + u = 1.44$; $R(C=C) = R_o - u = 1.36$; $R(C\equiv C) = 1.21$. Bond angles: $2\pi/3$ at sp^2 carbons, π at sp carbons. Variations of $\lesssim 0.02A$ and $\lesssim 5°$ in bond lengths and angles, respectively, are unimportant here; X-ray and quantum chemical data give such refinements.

quantum cell models as approximations to the Schrödinger equation
and as phenomenological or effective Hamiltonians for elementary excitations in solids. Quantum cell models consequently bridge, however imperfectly, experimental observations and basic theory by
focusing on the frontier orbitals ϕ_p that dominate the states near
the Fermi energy, ε_F.

Theoretical models have many purposes. They should ideally be
simple yet general, combining insights with quantitative predictions,
and capturing essential features in analytical form. Particles-in-box
(free electron) models for metals or polymers and harmonic oscillators for small displacements illustrate the wonderfully wide scope
of simple exact solutions. Quantitative analysis usually requires
some corrections, however, that are less tractable and general. Boxes,
harmonic oscillators and Hückel models will remain important without
ruling out, respectively, atomic centers, anharmonic terms, or electronic correlations; they identify properties that are insensitive to corrections. Quantum cell models have important advantages even when
the simplicity of Hückel solutions is lost.

The restriction to a single ϕ_p per site naturally unites wide and
narrow band problems, as discussed [22] in connection with
localization in Hubbard models. It also unites molecular PPP problems with models for charge-transfer and ion-radical organic solids,
[21] where ϕ_p is the HOMO of a planar π-donor (D) or the LUMO
of a planar π-acceptor (A). Many other solid-state models involve
localized states and/or s-bands that are readily translated into one
ψ_p per site. These models have widely different transfer integrals
$t(R)$ in Eq. (2), on-site interaction U_p in Eq. (6) and intersite interactions $V_{pp'}$ in Eq. (8). Strong on-site electron-phonon coupling
can lead to negative (attractive) U_p, and such models may be appropriate for dangling bonds [23] in Si. The direct evaluation of
the parameters in Table 1 is simpler for π-electrons, where the ϕ_p
are essentially atomic orbitals and the integrals are large. But
the parameter choices are separate issues. As discussed by Soos
and Klein [21], all intrasite (core) electronic and nuclear relaxation
processes may be rigorously included phenomenologically when only
the ground electronic states for $n_p = 0$, 1, or 2 are considered.
These are just the approximations leading to the PPP model for conjugated molecules.

As $U_p \rightarrow \infty$, doubly-occupied ϕ_p are excluded. The atomic limit
[24] in physics amounts to retaining the most covalent diagrams in
chemistry. Now Hubbard models reduce to noninteracting *spinless*
electrons, simple exact results are again possible, and there is a
clean separation in one-dimensional systems between charge and
spin degrees of freedom. Regular half-filled arrays of A^- or D^+
ion-radicals are in fact semiconductors [21], and thus illustrative of
narrow rather than metallic bands in Mott's famous analysis [25] of
metal-insulator transitions. Such a choice is not forced for either
PA or PDA, however, since alternating bond lengths in Figs. 1 and

2 lead to a semiconductor even in the absence of correlations. For $U_e = U_p - V_{p,p+1} \gg t_p$, the spin degrees of freedom are described by a Heisenberg exchange Hamiltonian [26],

$$H_{spin} = \sum_p 2J_p \, (\vec{s}_p \cdot \vec{s}_{p+1} - 1/4) \tag{10}$$

with $J_p = 2t_p^2/U_e$. Virtual electron-hole pairs always lead to anti-ferromagnetic exchange [27]. The network is one-dimensional whenever the transfers in Eq. (4) occur in strands. Exchange networks [28] have been extensively studied in connection with magnetic insulators. Any $s_p = 1/2$ system is immediately recognized to be a limiting case of a quantum cell model, as are larger s_p in slightly more general terms.

Valence-bond (VB) diagrams have long been used in organic chemistry [29] as a many-electron representation of both ground and excited states. The Kekulé structure in Fig. 1, for instance, implies a Heitler-London covalent bond between the π-electrons in double bonds. An ion-pair involving a carbocation C^+ and a carbanion C^- can readily be imagined. VB diagrams immediately give the occupation number $n_p = 0$, 1, or 2 for the $2p_z$ orbital at atom p. As shown in Fig. 3 for benzene, there may be more than one pairing and the equivalence of all bond lengths is understood in terms of resonance. The point here is that, within the restriction of one ϕ_p per site, the localized basis of C^+, C^-, and covalent sites with $n_p = 1$ is complete [30] and uniquely related to VB diagrams with arbitrary total spin $S \leqslant N_e/2$ and charge $q = N - N_e$. Every diagram can be expanded in terms of N_e-electron Slater determinants, the conventional many-electron basis. Thus VB diagrams form a convenient and rigorous many-electron basis for quantum cell models and provide a quantitative basis for Pauling's familiar notion of admixing various states.

The half-filled case of one-electron per ϕ_p is typical for conjugated polymers and for many other solid-state applications. Some

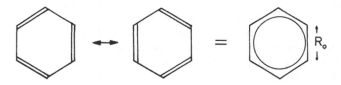

Fig. 3   Kekulé structures and resonance in benzene.

powerful theorems then hold for alternant systems [10,31], in which ε_p in Eq. (1) and U_p in Eq. (6) are independent of p, only nearest-neighbor t_p occur in Eq. (2), intersite interactions $V_{pp'}$ in Eq. (8) are spin-independent, and the sites can be decomposed into two sets such that all bonds (finite t_p) involve one site from each set. Half-filled alternant cell models have electron-hole symmetry [32,33] associated with interchanging particles and holes; other names [34] for this feature of half-filled models are pseudo-parity, charge conjugation, and alternancy symmetry. The exact cation and anion energies, moreover, are related by the additive constant, $-2e + U$, for a doubly occupied site.

E. Electron-Electron Correlations

Many approximation schemes can be applied to Hubbard or PPP models. We consider the ground state $|g>$ of a 2n-electron, N-site system. The Hückel Hamiltonian (4) leads to N eigenvalues λ_k, $k = 1, 2, \ldots, N$, and eigenfunctions $\psi_k(\vec{r})$. The Slater determinant

$$|g> = \det |\psi_1 \alpha_1 \psi_1 \beta_2 \cdots \psi_n \beta_{2n}| = \prod_{k \leqslant n} a_{k\alpha}^+ a_{k\beta}^+ |0> \qquad (11)$$

corresponds to doubly filling the n lowest-energy orbitals. The filling is unique for nondegenerate $|g>$ and $|0>$ is the vacuum.

The ψ_k or $a_{k\sigma}^+$ are linear combinations of ϕ_p or $a_{p\sigma}^+$, respectively. In self-consistent-field (SCF) theory, the ϕ_p and other valence orbitals are optimized and the Hartree-Fock (HF) limit is the lowest-energy function of the type (11). Quantum cell models usually have fixed ϕ_p and thus resemble minimum-basis HF in which Eq. (11) merely gives the best that can be done with the given ϕ_p.

SCF treatments of radicals lead to unrestricted Hartree-Fock (UHF) solutions [35] that retain the single-determinantal form of Eq. (11), but have different ψ_k for spins α and β. The resulting $|g>$ is no longer an eigenstate of S. More complicated SCF treatments [36] in which S is conserved can be developed in terms of linear combinations of Slater determinants. Canonical transformations [11] for solid-state problems, as illustrated in the BCS theory of superconductivity, are other examples of one-electron schemes that go beyond a single determinant of plane waves with $k \leqslant k_F$. The traditional definition of the correlation energy as the difference between the exact and HF solution is not particularly useful for quantum cell models, since restricted ϕ_p spoil the HF limit, other SCF possibilities occur for open shells, and accurate spectroscopic or other experimental results are not relevant to model Hamiltonians.

The finite basis of N_e electrons and N orbitals ϕ_p leads to a large but finite number of Slater determinants. The resulting secular determinant can be diagonalized. Configuration-interaction (CI) schemes are based on linear combinations of Slater determinants, with the largest coefficient usually associated with $|g>$, the dominant configuration. We may use the Hückel MOs, ψ_k, in Eq. (11) or SCF MOs whose convergence may be more rapid. One difficulty with Slater determinants is that only S_z is conserved, on taking p electrons with spin α and N_e-p with spin β. Linear combinations of Slater determinants are needed for eigenstates of S^2.

The localized basis of VB diagrams automatically conserves S [30] and provides an alternative method for exact solutions. N_e = 2n electrons on N sites with total spin S = 1, 2, . . ., n lead to [31]

$$P_s \ (2n,N) = \frac{2S + 1}{N + 1} \begin{bmatrix} N + 1 \\ n + 1 + s \end{bmatrix} \begin{bmatrix} N + 1 \\ n - s \end{bmatrix} \qquad (12)$$

linearly independent VB diagrams. The binomial coefficients in Eq. (12) describe an exponential increase with N that more than offsets reductions from spatial or electron-hole symmetry. Nevertheless, exact solutions [31,37] for quantum cell models with $N = N_e \leqslant 14$ are currently feasible this way. Extrapolations or scaling arguments are needed for larger N. Monte Carlo methods [38] provide ground-state information up to N = 50.

The eigenstates are now linear combinations of VB diagrams and may be described in terms of dominant Kekulé structures for PA in Fig. 1 or PDA in Fig. 2. Just as full CI involves many configurations beyond $|g>$ in Eq. (11), however, the exact states are highly admixed [31] combinations of VB diagrams. Although $|g>$ in Eq. (11) and Kekulé structures like Figs. 1 and 2 are quite different, both contain important information about the π-system and are interchangeably used in different contexts. The atomic limit for N_e = N corresponds to purely covalent diagrams, which also give a complete basis for Eq. (10). For $N_e \neq N$, doubly occupied C^- sites are minimized for large U.

II. ELECTRONIC STRUCTURE OF IDEALIZED SINGLE STRANDS

A. Bond Alternation in Cyclic Polyenes

The familiar sp^2 and sp hybridization at double and triple bonds, respectively, in Figs. 1 and 2 suggests that molecular bond lengths and angles are preserved in conjugated polymers. Both experiment and theory support such assignments for PA, the PDA backbone,

and other conjugated polymers. Cyclic polyenes $C_N H_N$ with even
N, N_e = N, and alternating transfer integrals $t(1 + \delta)$ in Eq. (2)
represent idealized PA strands as N → ∞ and form separate N = 4n
and N = 4n + 2 molecular series associated, respectively, with anti-
aromaticity and aromaticity [10,37]. Cyclobutadiene (N = 4) is
rectangular, or alternating, with partial double and single bonds
[39]. Benzene (N = 6, Fig. 3) has equal bond lengths R_0 = 1397 A
that serve as the reference for half π-bonds. The extra stability
of aromatic systems has long been recognized and is illustrated by
annulenes [10] (N_e = N = 4n + 2, n = 3,4,5,6). The question of
regular (δ = 0) vs. alternating (δ ≠ 0) bond lengths for long
polyenes has been studied repeatedly, both experimentally and
theoretically, before being settled in favor of alternation [15,40].
More recent theoretical work is cited in Section III.B.

Cyclic polyenes $C_{2n} H_{2n}$ with N_e = 2n and D_{nh} symmetry are
particularly simple. Symmetry fixes completely the proper linear
combinations of ϕ_p. The Hückel MOs thus coincide with HF orbitals
for Hubbard, PPP, or other potentials. Coulson [41] first treated
linear polyenes systematically. The one-electron energies of alternating Hückel rings are

$$\lambda_k (\delta) = \pm 2|t| (\cos^2 k + \delta^2 \sin^2 k)^{1/2} \tag{13}$$

on taking c_p = 0 in Eq. (1). The Hückel energies are shown in Fig.
4 for N = 8 and 10, as well as for N → ∞. The size dependence is
entirely contained in the N/2 wavevectors of the dimerized chain

$$k = 0, \pm \frac{\pi}{2n + 1}, \pm \frac{2\pi}{2n + 1}, \cdots, \pm \frac{n\pi}{2n + 1} \quad N = 4n + 2$$

$$k = 0, \pm \frac{\pi}{2n}, \cdots, \pm \frac{(n - 1)\pi}{2n}, \frac{\pi}{2} \quad N = 4n \tag{14}$$

in the first Brillouin zone. The N/2 lowest (bonding) levels are
filled in Fig. 4 in the ground state |g>, which has the form of
Eq. (11). The assignment is unique for δ > 0, but regular (δ = 0)
4n rings have orbitally degenerate |g>. The bandwidth in Fig. 4
is W = 4|t|, independent of δ. The ground state energy per site
in units of W is

$$\bar{\lambda}(N,\delta) = \frac{E(N,\delta)}{WN} = -\frac{1}{N} \sum_{k \text{ filled}} (\cos^2 k + \delta^2 \sin^2 k)^{1/2}$$

$$= -\frac{1}{\pi} E(\sqrt{1 - \delta^2}) \quad N \to \infty \tag{15}$$

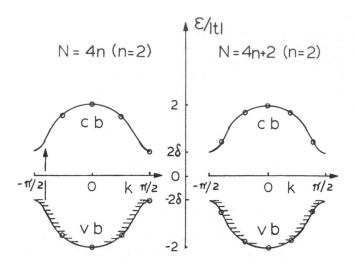

Fig. 4   Schematic band structure, Eq. (13), of alternating Hückel rings. The wavevectors k in Eq. (14) are indicated by open dots for N = 8 (left), 10 (right), and ∞ (solid line). The filled valence band is suggested by hatch marks and the Fermi energy is chosen at $\varepsilon_F = 0$.

The sum is over the k values in Eq. (14) and E is the complete elliptic integral of the second kind. The half-filled finite band is unstable to dimerization [15], as recognized by Peierls [42]. A harmonic lattice or σ-compressibility of $K\delta^2/2$, with $t\delta = \alpha u$ for alternation ±u in Eq. (3), is always overcome at sufficiently small δ by the $\delta^2 \ln|\delta|$ term in the expansion of E.

The Jahn-Teller theorem ensures the instability of regular 4n Hückel rings. To lowest order the electronic energy per site is reduced by $-a(n)|\delta|$ on alternation, which clearly wins against a harmonic potential $K\delta^2/2$. This rationalizes rectangular C_4H_4 at the Hückel level. The coefficient a(n) vanishes as n → ∞, however, since only the band edge (k = π/2) states are split linearly while those near k = 0 are not shifted at all. There is no degeneracy in regular 4n + 2 Hückel rings, whose energy per site goes as $-b(n)\delta^2$. The uniform bond lengths of benzene require K > 2b(1) and finite 4n + 2 rings are conditionally stable. But b(n) increases with n and, as already noted, diverges logarithmically in the infinite polyene.

The sum in Eq. (15) is elementary for $\delta = 0$. We obtain

$$\frac{E(N,0)}{WN} = -\frac{1}{N} \cot \frac{\pi}{N} \simeq \frac{-1}{\pi} + \frac{\pi}{3N^2} \qquad N = 4n$$

$$\qquad\qquad = -\frac{1}{N} \csc \frac{\pi}{N} \simeq \frac{-1}{\pi} - \frac{\pi}{6N^2} \qquad N = 4n + 2$$

$$(16)$$

The $N \to \infty$ limit is approached from above by the $4n$ series, from below by the $4n + 2$ series. The former is destabilized, the latter stabilized, relative to the infinite regular chain. For regular Hückel rings, any $4n+2$-electron molecule or ion has a nondegenerate ground state, while $4n$-electron species are degenerate and subject to Jahn-Teller distortions.

The dimer limit $\delta \to 1$ reduces both $4n$ and $4n + 2$ polyenes to $N/2$ noninteracting ethylenes. There is no dispersion in $\lambda_k(1) = 2|t|$ and the delocalized MO representation can equally well be written as a localized bonding-antibonding pair for each dimer. The sum in Eq. (15) is now trival and $E(N,1)/WN$ reduces to -0.5 for any given N. The size dependence disappears as required in this limit. For intermediate $t\delta = \alpha u$ in Fig. 4, the size dependence of the ground-state energy per site reflects the band-edge region. $k \sim \pi/2$. The spacing of the wavevectors in Eq. (14) is $2\pi/N$. Although $4n$ and $4n + 2$ alternating polyene rings still approach the $N \to \infty$ limit from above and below, respectively, the differences in $E(N,\delta)/WN$ become negligible for $N\delta > 2\pi$. The same conclusion may be reached by noting that the minimum excitation energy 4δ in Fig. 4 sets the scale $\zeta \sim \delta^{-1}$ for the correlation length and finite size effects become unimportant for $N > \zeta$.

The ground-state degeneracy of half-filled regular $4n$ rings is lifted by correlations. Exact results [43] are restricted to the infinite regular Hubbard chain, whose excitation spectrum is treated by Ovchinnikov [44], and to finite alternating Hubbard or PPP rings In the atomic limit $(U \gg |t|)$, the Heisenberg antiferromagnetic chain in Eq. (10) has been solved [45] up to $N = 26$ for alternating exchanges $J(1 \pm \delta)$ and cyclic boundary conditions. Bonner and Blöte [46] have carefully discussed various approaches to the spin-Peierls instability in alternating spin-1/2 Heisenberg antiferromagnets, a related but simpler problem. Regular even rings converge nearly as N^{-2} from below, with both $4n$ and $4n + 2$ systems falling on the same line. The correlation gap $\sim U_e$ at $\delta = 0$ evidently removes any trace of the $4n$ degeneracy. Small but finite correlations, on the other hand, preserve the different convergence of

E(N,0)/WN in Eq. (16). The natural crossover to common behavior is expected when finite-chain effects of order $4|t|/N$ become smaller than the correlation gap of order U_e. In $U = 4|t|$ regular Hubbard rings, for example, the ground state energies per site [31] go as $\bar{\lambda}(12) < \bar{\lambda}(8) < \bar{\lambda}(\infty)$ and clearly imply a subsequent increase with N as found in the 4n + 2 series. Similarly, molecular PPP parameters indicate [37] that regular 4n rings become aromatic for $n \geqslant 4$ and converge from below with the 4n + 2 rings. Larger U in Eq. (9) shifts the crossover to smaller N as expected.

The question of alternation in half-filled Hubbard rings has been discussed by Dixit and Mazumdar [47]. SCF [48] and perturbative analyses [49] generally show the tendency toward alternation to decrease with increasing U, while exact analysis of finite rings [47] indicates the opposite up to $U \sim 4|t|$. The initial enhancement of dimerization with U in Hubbard models has been supported by such nonperturbative techniques as quantum Monte Carlo [50], the Gutzwiller ansatz [51] and renormalization group theory [52]. Since Hubbard models are restricted to on-site interactions, the potential is independent of δ. Potential contributions in PPP models further increase the tendency to dimerize [31]. On the other hand, the partial inclusion of terms beyond the ZDO approximation for δ-function potentials leads to the opposite conclusion [53]. The proper choice of models, potentials, and solutions are all intertwined in such theoretical debates. In the present context, we emphasize that the observed bond-length alternation in conjugated polymers by no means fixes uniquely the e-p coupling constant α in Eq. (3) or the σ-compressibility K in Table 1.

B. Electron-Phonon Coupling: Topological Solitons

We turn next to linear polyenes $C_N H_{N+2}$ and consider in Fig. 5 radicals with odd $N = N_e$ and ions with odd N and even $N_e = N \pm 1$. Linear polyenes are alternant [10], with paired MOs in either Hückel or SCF theory when e_p in Eq. (1) and U_p in Eq. (6) are independent of p, the arbitrary t_p in Eq. (2) are restricted to neighbors, and the potential $V(p,p')$ in Eq. (8) is spin-independent. It follows that for the HOMO for a radical (odd $N = N_e$) is precisely at e = 0, the reference energy in Fig. 6, and is a nonbonding orbital with nodes at odd sites on taking the origin at the center. The unpaired electron in Fig. 5a is consequently delocalized on even sites, as are the charges for $N_e = N \pm 1$ in Fig. 5b. The nodal structure of the HOMO is fixed in all three cases by the alternancy symmetry. McLachlan [32] realized long ago that the exact, or full CI, excitation energies of the anion and cation are consequently identical within the PPP model and account for similar hyperfine shifts in alternant anion and cation radicals. Charge delocalization

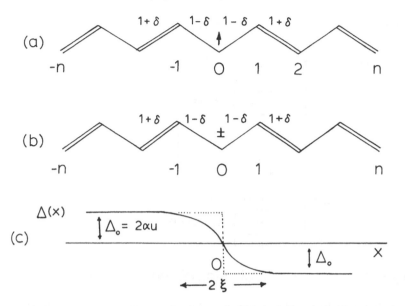

Fig. 5   (a) Schematic representation of a neutral spin-1/2 soliton at the origin; (b) a charged spinless soliton at the origin; (c) alternation crossover in continuum limit, with the dashed line representing the zero-width solitons in (a) or (b) and the $\xi > 0$ solid line giving the lowest-energy bond-length distribution.

in carbocations has been hotly debated [54] in terms of classical (localized) vs. nonclassical (delocalized) charge distributions. Electron-phonon coupling in conjugated polymers raises similar issues.

Pople and Walmsley [55] recognized the connection between localized nonbonding states, or alternation defects, and the weak paramagnetism of conjugated networks like graphite. Similar but less prescient conclusions were reached [56] for Hubbard models. Su, Schrieffer, and Heeger [13] (SSH) decisively extended the analysis of nonbonded states by solving quantitatively the infinite Hückel chain with linear e-p coupling. The SSH Hamiltonian is

$$H_{SSH} = H_t + \sum_p K u_p^2/2 \tag{17}$$

Here $u_p = r_{p+1} - r_p - R_o$ is the deviation of the pth bond length from R_o, the reference for half π-bonds that minimizes the (harmonic)

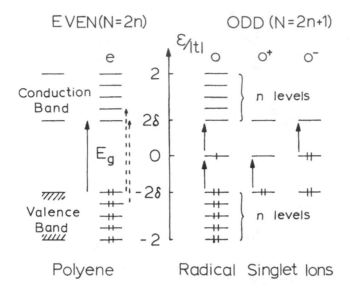

Fig. 6   Schematic band structure and lowest-energy $\pi-\pi^*$ excitation of alternating even and odd Hückel chains. Odd-length chains have a midgap level that is singly occupied in the radical, empty in the cation, and doubly filled in the anion.

lattice energy. The $t(u_p)$ in Eq. (4) are expanded according to Eq. (3) and the overall chain length is held constant. Within the adiabatic approximation, the u_p are optimized separately for various electronic states. The SSH model summarized below has been extensively [1] discussed, generalized, and related to field theories. It provides a detailed picture of elementary excitations, neutral and charged solitons, that rationalize many experimental observations and suggest additional tests. Current discussions about its quantitative merits indicate just how many new avenues it has opened up. Like free-electron theories of metals or Hückel models of molecules, however, the essence of the SSH model does not lie in quantitative comparisons.

   The SSH ground state for *trans*-PA is doubly degenerate, since double and single bonds may be interchanged for an infinite polyene in Fig. 1. The bond lengths alternate as $u_p = -(-1)^p u$; the transfer integrals alternate as $t_p = t(1 \pm \delta)$ with $t\delta = \alpha u$. The parameters t, α, and K are specified from the band width $W = 4t \sim 10$ ev, the optical gap $E_g = 4|t|\delta \sim 1.4$ ev, and a bond-length difference of $2u \sim 0.08$ A. The benzene ground state in Fig. 3 is nondegenerate, with uniform bond lengths, $u_p = 0$ for all p, and resonance

among the two Kekulé and many other VB diagrams. The two
valence tautomers of rectangular cyclobutadiene readily interconvert
[57] by tunneling through a barrier of about 0.5 ev ($\sim$10 kcal/mol).
Direct interconversion of the ground states does not occur in infinite polyenes.

SSH consider the domain wall, or topological soliton, between regions with opposite alternation. For example, terminal double bonds
in Fig. 5 for odd-length polyenes ensure an alternation crossover.
The indicated crossover at the central atom is sharp and corresponds
to a soliton with zero width. Short radicals or ions have uniform
bond lengths, however, as expected from resonance arguments. An
infinite radical may consequently be expected to have reduced bond-
length alternation over some distance 2ξ, as sketched in Fig. 5c,
between regions with uniform but opposite alternation. The width
2ξ is completely determined within the SSH model by minimizing the
total energy as shown in Fig. 7 for several choices of E_g and soliton
width ξ. The minimum of the $E_g = 1.4$ ev curve fixes the soliton
width of $2\xi \sim 15R_0$. Since e-e interactions are omitted, either the

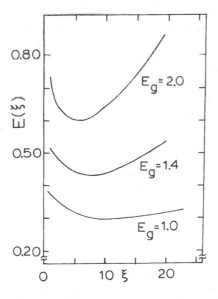

Fig. 7   Energy, in ev, of a soliton in the SSH model for various
optical gaps E_g and widths in Eq. (18). The minimum for
$E_g = 1.4$ is around $\xi = 7$. In each case, $2E(\xi)$ is less than E_g,
the minimum energy for an electron-hole pair in the rigid lattice.
(From Ref. 13.)

spin or charge in Fig. 5 is delocalized over some 15 carbons. They
are nonclassical in the physical-organic sense [54]. The neutral
soliton S in Fig. 5a has spin-1/2, while the charged solitons S^+ or
S^- have singlet ground states, thus showing reversed spin-charged
relations in the field-theory sense [58]. The order parameter
$(-1)^p u_p$ in Fig. 5c conveniently summarizes the shape of the soliton.
The SSH result for u_p is

$$u_p = R_{p+1} - R_p - R_0 = (-1)^p \frac{u}{2} [\tanh p/\xi + \tanh (p+1)/\xi] \quad (18)$$

for a soliton centered at the origin. The functional form in Eq. (18)
holds for other models as well. The width ξ varies [59–61] and
need not be the same for S, S^+, or S^- on including e-e contributions
or on going beyond π-electron theory [60].

Alternation crossovers are clearly and quantitatively described
within the SSH model. Solitons must be created in pairs on infinite
PA strands with even N. They may move along infinite chains,
thereby interconverting partial single and double bonds. Their role
as elementary excitations requires a pair of solitons to have lower
energy than the minimum electron-hole excitation, $E_g = 4|t|\delta$ in Fig.
4, for a rigid band. SSH explicitly show this for the discrete lattice. The soliton energy E_S in Fig. 7 is lower than $E_g/2$. Most of
the lowering from $E_g/2$ occurs even for sharp crossovers ($\xi = 0$),
with modest additional stabilization due to delocalization ($\xi > 0$).
The continuum limit sketched in Section II.C leads to simple analytical results. The alternancy symmetry of the SSH model (17) insures
that the soliton states in Fig. 6 are exactly at the middle of the
gap. Their stabilization is due to infinitesimal changes of all band
levels. The midgap state is singly occupied for S, empty for S^+,
and doubly filled for S^-. The SSH model thus gives optical absorptions at $E_g/2$ due to either neutral or charge solitons. These predictions are changed by e-e interactions as discussed in Section II.D.

The SSH midgap state is shown in Fig. 8 in terms of the coefficients c_p of the ϕ_p for a soliton centered at the origin in Fig. 5.
The spin is confined to the crossover region. By contrast, if we
take $u_p = 0$ for all p in a polyene radical of length $4n + 1$, the
Hückel result is $c_p^2 = (2n + 1)^{-1}$ at the $2n + 1$ even sites in Fig. 5
and $c_p = 0$ at the others. The SSH wavefunction in Fig. 8 corresponds
to a spin-density wave (SDW) for a neutral soliton and to a charge-
density wave (CDW) for charged solitons.

The unusual spin resonance of odd-alternant radicals has been
discussed by McConnell, Carrington, and others [62]. Instead of
vanishing unpaired-electron densities where the nonbonding MO has
nodes, the opposite spin polarization (negative spin density) is
found. One-electron theory, even at the SCF level, necessarily

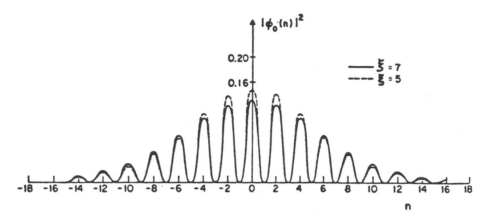

Fig. 8   The nonbonding MO, $\phi_0(n)$, in the SSH model for crossovers centered at the origin and width $\xi = 7$ and 5 in Eq. (18);
$\phi_0(n)$ has nodes at odd n.   (From Ref. 13.)

misses this effect.   The allyl radical (C_3H_5, $R_1 = R_2 = R_0$) always has
a node at C_2 for the nonbonding electron.   Resonance between the
purely covalent diagrams $\overset{\bullet}{C}$—C=C and C=C—$\overset{\bullet}{C}$ leads to negative spin
density $\rho_2 = -1/3$ and larger $\rho_1 = \rho_3 = 2/3$.   As so often, experiment is about in between [63].   Negative spin densities in PA are
observed in ENDOR, as discussed by Dalton, Thomann, and coworkers [64,65].   They emphasize that such observations imply e-e
correlations, require essentially constant positive and negative amplitudes ($\rho-/\rho+ \sim .3$) rather than the amplitudes in Fig. 8, and
imply $\xi \sim 25$ rather than $\xi \sim 7$ for static solitons.   Soliton motion
does not average [66] the positive and negative spin densities, however, and can be chosen to be consistent with any ξ.   Restricting
soliton motion within the infinite chain is readily understood in
terms of finite odd-length polyenes discussed in Section III.D.   That
e-e contributions are again seen for nonbonding orbitals is no sur-
prise.   Quantitative implications about the nature and width of the
paramagnetic center, on the other hand, are still open [1,5].

A polaron sketched in Fig. 9a is another excitation and involves
a local weakening of the alternation without producing a crossover.
Cation or anion radicals of even polyenes may, for example, be
described as bound pairs of neutral and charged solitons.   The
shape of polarons is also given analytically within the continuum
version of the SSH model in the next Section.   The PA dimerization
and its changes on electronic excitation involve optical phonons.   In
PDAs, polaron coupling to acoustic modes has been proposed by
Wilson and coworkers [67] to explain the extraordinarily long mean

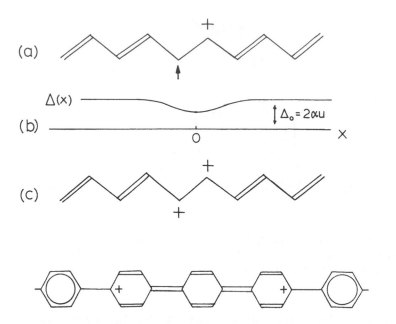

Fig. 9   (a) Schematic representation of a radical ion or polaron;
(b) alternation profile of polaron in the continuum limit, as given
by Eq. (24); (c) bipolarons in PA and in polyparaphenylene.

free paths found in photoconduction. The observation of conductivity without paramagnetism in polypyrrole, polythiophene, etc. has
been explained in terms of bipolarons [68], which formally amount
to pairing the two spins of two polarons in Fig. 9b and leaving
bound charges. Without pursuing the nature of these low-lying
states, we note that electron-phonon coupling is central and leads
to states below E_g.

C. Continuum Limit and Field Theory

The SSH model focuses on the nonbonding states in Fig. 6 of an
odd polyene or even numbers of such alternation crossovers in infinite chains. We have only sketched the *trans*-PA case of a degenerate ground state. The charge or spin associated with the
midgap state is delocalized over 2ξ sites, as understood by chemists
in terms of resonance among VB structures and by physicists as a
finite correlation length ξ at domain walls. The SSH description
includes the soliton wavefunction (Fig. 8), the lattice distortion u_p
in Eq. (18) and reversed spin-charge relations. Generalization to

nondegenerate ground states is straightforward. Soliton or polaron wavefunctions are zeroth-order states for considering e-e interactions.

Takayama, Lin-Liu, and Maki [69] (TLM), Brazovskii [70], and Krumhansl et al. [71] derived a continuum version of the SSH model, Eq. (17) with vanishing lattice spacing R_0. The TLM results summarized below agree well with SSH theory for features extending over $2\xi \sim 15R_0$. The regular chain energies, $\lambda_k(0) = 2t \cos k$ in Eq. (13) are modified near $k_F = \pi/2$ by lattice distortions. Expanding Eq. (13) about k_F leads to

$$\varepsilon(q) = \pm\hbar\, v_F\, q \tag{19}$$

for $q = |k - k_F|$ and Fermi velocity $v_F = 2|t|/\hbar$. Although $\varepsilon(q)$ in principle diverges, a typical field-theoretical cutoff is imposed to match the bandwidth $4|t|$ of the SSH or Hückel model. The TLM model retains linear e-p coupling in the adiabatic limit for noninteracting electrons.

The TLM model is equivalent to the N=2 Gross-Neveu model of relativistic quantum field theory. Campbell et al. [72] have explicitly sketched this connection and its implications to both areas. Techniques and concepts from field theories can be applied to solid-state problems relevant to conjugated polymers. One limitation, that phonon dynamics and quantum fluctuations can only be treated semiclassically, is related to the adiabatic approximation. Another limitation, that e-e interactions are ignored or handled perturbatively, is contrasted in Sections II.D with exact solutions to quantum cell models. The TLM limit of the SSH model leads to analytical results for alternation crossovers, as sketched in Figs. 5c or 9b, and thus clarifies the properties of solitons and polarons.

In the continuum model for infinite chains, there are two coupled electron fields for electrons moving to the right and left, respectively, or alternatively on the even and odd sites of the dimerized unit cell. The phonon field, or the gap order parameter Δ, represents the degree of bond alternation

$$\Delta(x=nR_0) = 4\alpha(-1)^n y_n \tag{20}$$

where α is the linear e-p coupling constant in Eq. (3) and y_n is the displacement of the nth site from $(n - 1)R_0$. In the ground state, $y_n = (-1)^n u/2$ leads to uniform alternation $R_0 \pm u$, Δ_0 is constant, and $2\Delta_0$ is the optical gap E_g. The TLM equations couple $\Delta(x)$ nonlinearly to the electron fields and are solved self-consistently. The ground state is uniformly dimerized and two-fold degenerate. Since the system is in one ground state or the other, the symmetry has been spontaneously broken. The magnitude of the gap parameter is [72]

$$\Delta_0 = We^{-1/2\lambda} \tag{21}$$

where $W = 4|t|$ is the bandwidth and $\lambda = 2\alpha^2/\pi Kt$ is the dimension-less e-p coupling constant for a harmonic lattice with force constant K. The dispersion relation replacing Eq. (19) is

$$\varepsilon(q) = \pm(\hbar^2 v_F^2 q^2 + \Delta_0^2)^{1/2} \tag{22}$$

and has a gap $2\Delta_0$ between the filled (bonding) and empty (anti-bonding) states. The behavior of Eq. (22) near $q = 0$ coincides with the λ_k spectrum in Fig. 4 near $k_F = \pi/2$ for $\Delta_0 = 2t\delta$. These are the states sought in the continuum limit to the discrete chain.

The lowest electronic excitation in Fig. 4 is expected at $2\Delta_0$. When e-p coupling generates the gap, however, this result is changed in an interesting and nontrivial way. The e-h pair can lower its energy by localizing around alternation crossovers. For the case of a degenerate ground state, two topological defects or solitons are found in field theories. The SCF solution for $\Delta(x)$ is

$$\Delta_s(x) = \Delta_0 \tanh(x - x_0)/\xi \tag{23}$$

for a soliton with delocalization length $\xi = WR_0/2\Delta_0$ centered at x_0. The alternation crossover has the form sketched in Fig. 5c. The explicit form of the midgap state is, aside from a normalization constant, $\mathrm{sech}(x - x_0)/\xi$ and is the enevelope of the SSH function in Fig. 8. The excitation energy of the soliton-antisoliton pair is $4\Delta_0/\pi$, less than $2\Delta_0$, and in good agreement with the SSH results E_S in Fig. 7. Topological considerations also regain the reversed spin-charge relations and the constraint to creating soliton-antisoliton pairs.

Adding or removing a single electron in the infinite strand leads formally to an ion radical, as sketched in Fig. 9a, and costs $2|t|\delta = \Delta_0$ relative to $\varepsilon_F = 0$ in Fig. 4. Once again, e-p coupling changes the band result because the spin and charge deform the lattice locally. SCF analysis of the TLM model for a polaron centered at $x = 0$ leads to [72]

$$\Delta_p(x) = \Delta_0 - k_0 \hbar v_F [\tanh k_0(x + x_0) - \tanh k_0(x - x_0)] \tag{24}$$

with x_0 given by $\tanh 2k_0 x_0 = \hbar k_0 v_F/\Delta_0$ and $\hbar k_0 v_F = \Delta_0/2^{1/2}$ for electron or hole polarons. The $\Delta_p(x)$ curve in Fig. 9b shows the alternation to be reduced by not reversed in the region of the charge and spin. Increasing the separation $2x_0$ leads to two reversals at $\pm x_0$, as suggested by Eq. (23). The polaron energy is $8^{1/2}\Delta_0/\pi$,

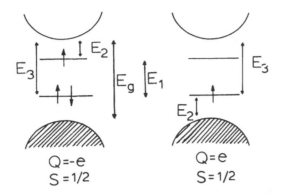

Fig. 10   Schematic representation of spin-charge relations, energy
levels, and subgap excitation of polarons in the continuum limit.
(From Ref. 72.)

which is less than Δ_0, and also leads to localized states in the gap as
as summarized in Fig. 10.

The position, shape, and transition probabilities for vertical transitions involving band and polaron levels in Fig. 10 are completely
specified [73] within the TLM model. Electron-hole symmetry leads
to identical spectra for polarons with positive and negative charge
even in the presence of spin-independent $V(p,p')$ in Eq. (8); but
neither spectrum is then known. Three transitions below E_g are
indicated in Fig. 10; E_1 is a δ-function transition between the gap
states, E_2 is a strong band-edge absorption and E_3 is a weak high-
energy feature. Similar results for bipolarons are discussed by
Campbell et al. [74]. They emphasize that the relative intensities
of in-gap features are not readily reconciled with experiment.

We conclude with several extensions of the continuum model. The
cis-transoid ground state in Fig. 11a cannot be degenerate with interchanged single and double bonds, although it is not obvious
which has lower energy. The butatriene form of PDA in Fig. 11b,
by contrast, is well above that of Fig. 2 and is associated with
short segments in the initial stages of the solid-state photopolymerization [75]. Pairs of alternation reversals are consequently bound by
the unfavorable conjugation in between. Polarons and bipolarons in
Fig. 9 are the elementary excitations of conjugated polymers with
nondegenerate ground states. The continuum limit is geared to nearly
degenerate ground states, as in cis-PA, since the initial dispersion
Eq. (19) is that of a regular chain. We suppose there is an intrinsic
alternation $\Delta_e = 2t\delta_e$ in addition to the phonon contribution Δ in Eq.
(20) and simultaneously treat their coupling to the electron fields

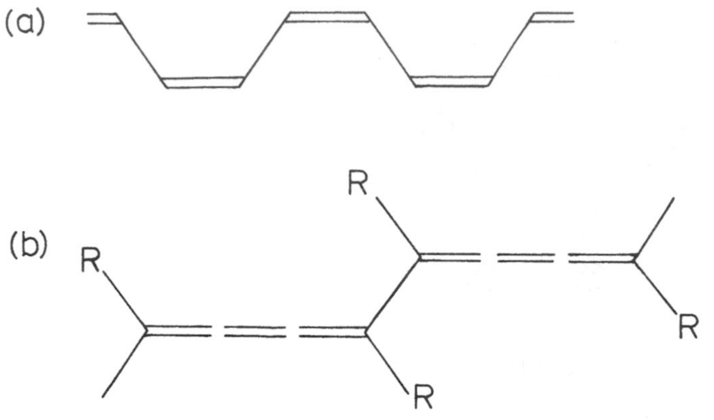

Fig. 11   (a) Schematic representation of cis-transoid PA; (b) Schematic representation of the butatriene structure of PDAs.

self-consistently. The ground-state alternation $\tilde{\Delta}_0$ then depends on Δ_e and λ, as well as on Δ_0 in Eq. (21). The principal result [72] is that alternation reversals now occur as bound pairs, since the intrinsic alternation Δ_e increases the energy linearly with their separation. Numerical rather than analytical results are found for the more general cases.

Conjugated heteronuclear polymers [76] $(AB)_x$ may also be treated in the continuum limit. The site energies e_p in Eq. (1) now alternate as $(-1)^n e$ for an ionization-potential difference of $2e$. Linear polyynes [77] or finite cumulenes [78] feature two orthogonal π-systems whose e-p coupling may be developed along TLM lines. Quarter-filled bands of ion-radical salts [79] provide still other possibilities within the continuum model. Field-theoretical approaches provide deep new insights about elementary excitation and their interconnections among quite diverse cases. These are somewhat tempered by the over-simplification of the model in neglecting e-e interactions and the experimental inaccessibility of the predicted excitations. All continuum results implicitly assume an infinite one-dimensional picture.

D. The 2^1A_g State and Optical Gap E_g

With increasing energy, the particle-in-box functions $\psi_n(x)$ are alternately even and odd about the center of the box. The same alternation between g and u MOs occur in regular Hückel chains or in all-trans even chains that preserve an inversion center. We consider for example, a PA fragment in Fig. 1 with 2n sites and alternating transfer integrals $t(1 + \delta)$, $t(1 - \delta)$, respectively. The Hückel eigenvalues $\lambda_k(\delta)$ in Eq. (13) are again found; the n values of the index k coincide for $\delta = 0$ with $k > 0$ for $4n + 2$ rings in Eq. (14) and are roots of a transcendental equation [15] for $\delta > 0$. As sketched in Fig. 6a, the MOs are nondegenerate and HOMO $\rightarrow$ LUMO excitation connecting a g,u pair is dipole-allowed. The lowest $\pi \rightarrow \pi^*$ transition to 1^1B_u is very intense in polyenes and other conjugated molecules. It evolves with increasing N into the optical gap E_g around 2 ev of conjugated polymers.

The pairing theorem for alternant hydrocarbons, or for alternant quantum cell models mentioned in Section I.D, ensures that the one-electron energies λ_k occur in pairs about the nonbonding state in Fig. 6. The paired MOs are related by a sign change for all expansion coefficients corresponding to every other site. This topological result, not restricted to one-dimensional arrays, holds for arbitrary t_p between neighboring sites and persists for SCF-MOs. The orbital pattern in Fig. 6 thus holds for *any* alternant linear, cyclic, polycyclic, and unsymmetrical π-system. The dotted arrows in Fig. 6 for $\varepsilon_1 + \varepsilon_2$ excitations are degenerate for alternant systems and are the second highest excitation for a half-filled band. The $\varepsilon_1 + \varepsilon_2$ transition is either g-g or u-u in trans polyenes and is consequently allowed in two-photon spectroscopy [80]. The 2^1A_g state has indeed been observed [81] in gas-phase trans polyenes with $n = 4,5$, and 6 double bonds, as summarized in Table 2. The striking result noted by Hudson and Kohler [82] is that 2^1A_g is below the dipole-allowed $\pi \rightarrow \pi^*$ excitation to 1^1B_u. The one-electron levels in Figs. 4 or 6 do not give the proper ordering of the excited states. Exact solution [31] to the PPP model with standard parameters restores the proper ordering in Table 2 and gives semiquantitative agreement with experiment. The ramifications of this fundamental e-e effect in excited states are discussed by Hudson et al. [83].

This correlation effect has a very general origin in half-filled quantum cell models [84]. The pairing theorem for alternant systems leads, at the one-electron level, to the degeneracy of $\varepsilon_1 + \varepsilon_2$ and of many other excitations in Fig. 6, while the E_g excitation is nondegenerate. On-site repulsions U_p in Hubbard models or Coulomb interactions in PPP models split the degeneracy and lead to an even and odd state under electron-hole symmetry. Dipole selection rules from the ground state require a change of electron-hole symmetry and are consequently restricted to one of the formerly degenerate partners. Increasing U in either Eq. (6) or Eq. (9) leads to the atomic limit and to a localized spin at every site, as discussed in Section I.D in connection with Heisenberg exchange, Eq. (10). The purely covalent VB diagrams, with $n_p = 1$ at every p, are a complete basis for Eq. (10) and *all* such states have the same electron-hole symmetry [31]. Thus no dipole-allowed transitions are possible within the spin manifold whose width is $J \sim 2t^2/U_e$. Physically, dipole transitions require charge redistribution, which cannot happen among purely covalent VB diagrams.

Table 2 Optical Gap Eg and Two-Photon Excitation Energy of Gas Phase Polyenes with N Carbons, N_e π-Electrons, and Alternation δ

N	N_e	δ	Eg (ev) PPP[d]	Eg (ev) Experimental	2^1A_g (ev) PPP[d]	2^1A_g (ev) Experimental
Molecules[a]						
8	8	0.07	4.561	4.40	3.775	3.59
10	10	0.07	4.234	4.02	3.404	3.10
12	12	0.07	4.001	3.65		2.73
Anions[b]						
5	6	0.0	3.456	3.42		
7	8	0.0	2.799	2.88		
Cations[c]						
5	4	0.0	3.456	3.13		
7	6	0.0	2.789	2.64		
9	8	0.0	2.343	2.25		
11	10	0.0	2.009	1.98		

[a]Experimental from Table 2, Ref. 83.
[b]Experimental from Ref. 93.
[c]Experimental from Ref. 90. $(CH_3)_2C(CH)_{N-2}C(CH_3)_2^+$ in concentrated H_2SO_4; slightly higher energies are reported for N = 7, 9, and 11 in heptafluorobutyric acid.
[d]The PPP results are for all-trans geometry and molecular parameters. (From Ref. 31.)

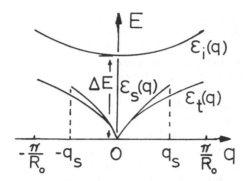

Fig. 12   Dispersion relation for the regular Hubbard chain.   The correlation gap $\Delta E = \varepsilon_i(0)$ for ionic excitations is given in Eq. (25). There is no gap for the covalent triplet $\varepsilon_t(q)$ or for covalent singlet, $\varepsilon_s(q)$, which has dispersion $-q_s \leqslant q \leqslant q_s$.   (From Ref. 44.)

Ovchinnikov [44] has found the exact excitations of the infinite regular Hubbard chain for arbitrary $U/|t|$.   The Hückel bandgap vanishes for $\delta = 0$ and $N \to \infty$.   The situation in Fig. 12 for $U > 0$ is qualitatively different.   Transitions to quasi-ionic states at $\varepsilon_i(q)$ have a minimum gap at $q = 0$ of

$$\Delta E \; \underset{\sim}{\,} \; U - 4t + \frac{8t^2}{U} \ln 2 + \dots \qquad\qquad U \gg t$$

$$\underset{\sim}{\,} \; \frac{8}{\pi} \, (Ut)^{1/2} \exp(-2\pi t/U) \qquad\qquad U < t$$

(25)

with $t = |t|$.   There is no gap at $q = 0$ for either quasihomopolar singlets at $\varepsilon_s(q)$, with $q \leqslant q_s$, or quasihomopolar triplets at $\varepsilon_t(q)$; the latter reduces for large $U/|t|$ to the triplet states of the regular Heisenberg chains found by des Cloizeaux and Pearson [85].   The half-filled regular Hubbard chain has gapless excitations leading to a finite magnetic susceptibility at $T = 0K$ and dipole-allowed excitations at $\Delta E > 0$.   The same features are expected in regular PPP or other chains that retain e-h symmetry.

Dimerization introduces a magnetic gap for $\varepsilon_s(q)$ and $\varepsilon_t(q)$ in Fig. 12 and has been discussed primarily for Heisenberg chains [45,46]. The 2^1A_g and many other covalent states of the infinite PPP chain may then be expected [31] to occur below the optical gap, E_g, defined by the lowest dipole-allowed transition for small δ.   For large δ and small U, the Hückel gap $4|t|\delta$ may be the lowest excited state.

While the functional dependence is not known, the general features
of the correlation crossover for 2^1A_g and 1^1B_u are clear.

Odd-alternant chains behave slightly differently. The pairing
theorem now requires a strictly non-bonding MO shown in Fig. 6
for any choice of the nearest-neighbor transfer integrals. If the
odd-length alternating chain has a C_2 axis, The MOs still alternate and there are two dipole-allowed transitions at $E_g/2$ in the
limit of large N. These are the midgap absorptions of neutral solitons
S in the SSH model. The related cation or anion in Fig. 6 has no
unpaired spin and a single dipole-allowed excitation at $E_g/2$. Correlations [86] affect the degenerate excitation of the neutral segment
o far more strongly than the nondegenerate excitation of o^+ or o^-.
The change in notation emphasizes that odd segments rather than alternation crossovers are examined. The optical excitations of o^+ or
o^- are identical due to electron-hole symmetry. The dipole-allowed
transition of o is shifted to higher energy, essentially to E_g. Correlations thus indicate that only S^+ or S^- excitations should be in
the region around $E_g/2$, and this expectation is confirmed [87] in a
trans-PA sample whose inadvertently produced holes are compensated
with NH_3 but whose paramagnetism still fixes the concentration of
neutral radicals.

The size dependence of $\pi \to \pi^*$ absorptions of conjugated molecules,
including many commercially important dyes, is among the oldest in
theoretical chemistry [10,88]. Such correlations were first studied
experimentally by comparing the lowest absorption peaks. Solvents
and substituents, even methyl and other nonconjugated groups, tend
to lower the absorption by increasing the polarizability [81,83]. Differential solubilities also hinder unambiguous comparisons for widely
different conjugation lengths. Longer polyenes extrapolate to a
finite gap E_g while longer cyanine dyes [89] $[(CH_3)_2N(CH)_{2n-1}N-$
$(CH_3)_2]^+$, or carbocations [90] appear to have vanishing optical gaps.
The lowest $\pi \to \pi^*$ maximum can be fitted to [91]

$$E_g(N) = E_g + A/N \qquad \text{(polyene)} \qquad (26)$$

$$= B/N \qquad \text{(cyanine dye)}$$

for N conjugated sites and different A, B constants. The extra-
polated value gap for polyenes, $E_g \sim 2.2$ ev, strongly supports the
giant-molecule approach to PA, PDA, or other conjugated polymers.
We do not consider here small (<0.5 ev) variations due to the
medium's polarizability, to interchain coupling, or to conformational
effects.

The different behavior of polyenes and cyanine dyes poses serious
difficulties for any one-electron treatment. The molecules have very
similar π-systems. Moreover, the polyene extrapolation in Eq. (26) is

close to E_g for PA, while the lower-energy features at 0.43 ev associated [92] with charged solitons fits extrapolations of ions. Any satisfactory π-electron theory must account for this qualitative difference, and e-e contributions do just that. The even polyenes in Table 2 have half-filled π-bands with $N_e = N$. Their optical gaps are shown in Fig. 13 for exact PPP solutions and standard parameters. Optical gaps for regular ($\delta = 0$) fragments are also shown in Fig. 13. The odd polyenes, or neutral radicals in Fig. 5a, now fall on the same curve as the even segments. As already noted, e-e contributions strongly raise the midgap absorption. Comparison of the $\delta = 0.07$ and 0.0 lines in Fig. 13 show E_g to depend primarily on e-e correlations rather than alternation.

The polyene ions and cyanine dyes, by contrast, have $N_e = N \pm 1$ rather than a half-filled π-band. Short o^+ and o^- segments are

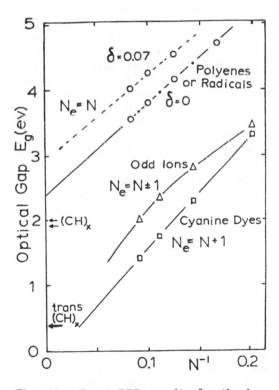

Fig. 13 Exact PPP results for the lowest π→π* excitation of alternating ($\delta = 0.07$) and regular ($\delta = 0.0$) polyenes, radicals, ions, and cyanine dyes with N conjugated centers and molecular parameters in Table 1. (From Ref. 31.)

expected to have regular bond lengths on chemical grounds and their
optical gaps in Fig. 13 are based on $\delta = 0$ and standard PPP parameters [31]. Gas-phase values [93] for the 5- and 7-carbon o⁻
fragments in Table 2 are within 0.2 ev of theory. Solution spectra
[90] on $(CH_3)_2C(CH)_{M-2}C(CH_3)_2^+$ for M = 5, 7, 9, and 11 in Table
2 also fit the PPP result and follow the cyanine extrapolation in Eq. (26)
Cyanine dyes $[(CH_3)_2N(CH)_{2n-1}N(CH_3)_2]^+$, are cations with $N_e =$
N + 1, since each terminal N contributes two π-electrons. They no
longer have electron-hole symmetry and require different site energies ε_N in Eq. (1) and on-site correlations U_N in Eq. (6). The
exact PPP results in Fig. 13 are for a regular chain with equal C-N
and C–C bond lengths. The $N_e = N + 1$ lines clearly extrapolate to
far lower values than the $N_e = N$ systems. As in the relative position of 2^1A_g and 1^1B_u, e-e correlations qualitatively change the
Hückel or SCF-MO result for E_g in $N_e = N$ and $N_e \neq N$ molecules.

We emphasize that Table 2 and Eq. (26) contain experimental
data. The PPP results are exact solutions to quantum cell models
that contain no adjustable parameters; the alternation δ is taken
from experiment and the microscopic parameters in Table 1 are based
on smaller molecules. The question of solid-state parameters is considered in Section III.C. The theoretical distinction between optical
excitations in $N_e = N$ and $N_e = N \pm 1$ could better be made in terms
of exact states [94,44] of the regular Hubbard model. But the
relative magnitude of U and the effect of alternation must then be
estimated for conjugated polymers. While the short, odd polyene ions
in Fig. 13 are nearly regular, for large N Peierls' theorem [42] and
the SSH analysis leads to an alternation crossover for charged soliton
between regions with uniform but opposite alternation. The 0.43 ev
PA absorption [92] is consequently expected at finite energy even
in the absence of pinning by the oppositely charged dopant [13].
In some ways the molecular data are thus preferable for the regular
infinite chain.

III. SELECTED APPLICATIONS TO CONJUGATED POLYMERS

A. Questions of Scale and Dimensionality

The SSH approach to infinite conjugated strands illustrates the basic
features of alternation crossovers arising from e-p interactions,
while PPP results for finite polyenes rationalize strong e-e effects.
Polyenes and PA are prototypical applications of these models. The
low-lying states of conjugated polymers pose a whole range of formi-
dable problems. Some will in retrospect be seen as small corrections
to single-strand models, but major reassessments are also possible.
Since we will discuss separate kinds of applications and comparisons

in the following section, it may be instructive to start with questions of length scales and dimensionalities. Such topics can rarely be treated simultaneously and typically raise different issues.

The electronic structure of conjugated polymers in Section III.B deals with local structure arising from many possible monomers. Idealized limiting structures [95] of three important conjugated polymers are sketched in Fig. 14. In each case, the upper form has the lower energy. Ionization of two nonadjacent rings leads to VB diagrams containing the lower structure in the region between the

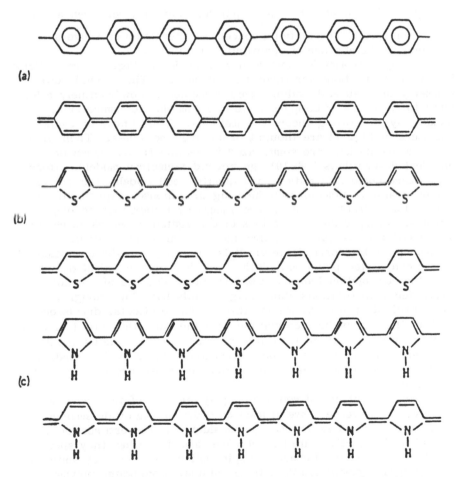

Fig. 14   Limiting VB structures of (a) polyparaphenylene (PPH) (b) polythiophene (PT), and (c) polypyrrole (PPy). In each case, the upper structure has lower energy. (From Ref. 95.)

charges, and no unpaired electrons. Such bipolarons [5,6,68] account
qualitatively for several in-gap states observed on doping or electrochemical oxidation and for the absence of paramagnetism. Their re-
semblance to the elementary bipolaron in Fig. 9c is clear. All these
polymers have nondegenerate ground states and their continuum
limits differ only in the choice of parameters. The structures in
Fig. 14 are, in fact, far more idealized than the PA or PDA backbone. The benzene rings in polyparaphenylene (PPH) are not co-
planar due to steric repulsion between ortho hydrogens. Every
other polypyrrole (PPy) or polythiophene (PT) ring is probably
flipped. Moreover, the ground-state alternation and e-p coupling
clearly involve several $a_{p\sigma}^{+}$, $a_{p\sigma}$ operators per unit cell even at the
π-electron level. The unifying concept of alternation crossovers
survives in the absence of quantitative details

The proper geometries and relative energies in Fig. 14 are difficult
but natural questions for theoretical chemists. The ground-state
geometry of small and medium-sized molecules is now routinely calculated to an accuracy comparable to experimental determinations,
usually by calibrating theoretical methods. Thus linear or cyclic
oligomers in Fig. 14 are thought of as large molecules. Their charge
and spin excitations are summarized in Section III.B. Quantum-
chemical descriptions [96–101] go beyond π-electron models, afford
more microscopic information than quantum cell models, but retain
single strands. Molecular engineering presupposes experimental control at the molecular level, where chemistry rather than topology
matters. Prospective applications of conjugated polymers as electroactive materials certainly require such molecular information.

Although PDAs form single crystals [2], they are the exception
among conjugated polymers. Either chemical or mechanical defects
limit the conjugation length. Conformational degrees of freedom
associated with rotations about single bonds have low enough energies
to be excited at ambient temperatures. These effects, discussed in
Section III.D, may be incompatible with the band-theoretical picture
of Section II for extended infinite strands. All the states should be
localized, for example, in a one-dimensional system with disorder, but
perhaps on a longer scale than alternation crossovers in conjugated
polymers.

Interchain contacts of conjugated polymers are of the van der Waals
variety. Transverse t' are at least an order of magnitude smaller
than $-t(R_0) = 2.40$ ev and may be considered to be perturbations.
Interchain interactions are nevertheless important when they spoil
purely one-dimensional charge or spin motion, or when they suppress
divergences in calculated densities of states. Conjugated polymers
are not necessarily one-dimensional, either, although such a description is practically synonymous with polymers. Polymer doping and

morphology are inherently three-dimensional and may involve inhomo-
geneties or domains on a scale comparable to the conjugation length.
Such questions are beyond our discussion of quantum cell models.

B. Quantum Chemistry of Conjugated Polymers

Electronic structure calculations have been so important for so long,
that even young fields defy comprehensive review. Recent reviews
by Silbey [96], Kertész [97], Whangbo [98], Brédas [99], and
Karpfen [100] cover diverse applications to conjugated polymers, and
many additional topics are discussed in recent conference proceedings [5–8]. We consider low-lying excited states and connections to
quantum cell models. Some oversimplifications below are deliberate,
others inadvertent.

The spin-density-wave (SDW), charge-density-wave (CDW) and
bond-order-wave (BOW) instabilities of the HF or UHF solution to
infinite Hubbard or PPP models have long been of interest [101–103].
The bond-length alternation of PA is correctly given, although care
is needed in UHF theory to avoid a regular chain. Fukutome and
Sasai [103] have applied SCF methods to discuss solitons and polarons
in PPP chains with linear e-p coupling. Their findings may be contrasted to the SSH results in Section II.B and II.C. Different crossover widths and negative spin densities are found. The PA optical
gap has also been carefully studied at the SCF level, with special
attention [104] to electron-hole stabilization of the 1^1B_u state. Such
quantum-chemical approaches to cell models illustrate different approximations. Some general conclusions [105] may be reached for spin-
independent potentials in Eq. (8).

Semiempirical schemes of the extended [106] Hückel or MNDO [107]
variety provide geometrical information beyond π-electron models.
Thus MO treatments of PPP models for conjugated polymers in Fig.
14 have been superceded by MNDO. The latter can be compared
with ab initio schemes, such as STO-3G, associated with Pople [108].
The goal is to demonstrate essentially HF accuracy in small molecules
at a fraction of the computational effort. The essential transferability
of chemical bonds is exploited. The procedure is then repeated for
as large a system as is practical, for either linear or cyclic boundary
conditions. The polypyrrole geometry based on MNDO and STO-3G
are compared [109] in Fig. 15a. The experimental values are for
a pyrrole monomer. The MNDO geometries [110] of the two limiting polythiophene structures are shown in Fig. 15b. The occurrence of local minima leads to unambiguous, and also unverifiable,
geometries for various limiting cases shown in Fig. 14. The lower
E_g for polyisothianaphthene (PITN) illustrates [111] molecular engineering by chemical modification of bond-length alternation. The rationale
for predicting such changes is consequently under scrutiny [110,111].

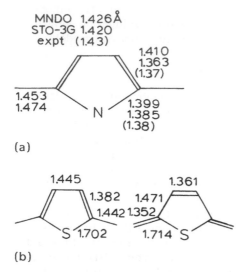

(a)

(b)

Fig. 15 (a) Comparison of MNDO, STO-3G, and experimental (monomer) bond lengths for pyrrole in PPy. The bond angles are comparable in both calculations. (From Ref. 109.) (b) MNDO bond lengths for the limiting VB structures of PT. (From Ref. 110.)

The relative energies of cis-transoid PA in Fig. 11, trans-cisoid PA, and all-trans PA in Fig. 1 differ [97] by less than 0.5 ev per C_2H_2 unit. The most recent results [112] place cis-transoid 0.1 ev above all trans, and trans-cisoid another 0.05–0.10 ev higher. Different orderings have also been reported. As in Madelung sums of electrostatic energies, care must be taken with boundary conditions and lattice sums [97]. The convergence of separate methods is in practice the simplest guide of reliability, and in this sense these calculations are computer experiments. Such quantum chemical results for the energy difference between nondegenerate ground states are future inputs for continuum SSH models.

STO-3G bond-length and charge-density changes in PPy are shown [113] in Fig. 16 on adding Na atoms above the central pyrrole rings. The limiting cases of undoped and 50% doping are modeled. There is almost complete (0.94 e) electron transfer from each Na, which is by no means the case for all dopants [99]. The band structure and density of states, which in general are also g ven b extended Hückel methods, are found in both limits. Na-doping of PPy is interpreted in terms of bipolarons, except at very low Na concentrations. Such microscopic information about doping has direct implication for bond-order reversals pinned at charged impurities. Pinning potentials V_i m y be estimated [13] but are hardly known accurately.

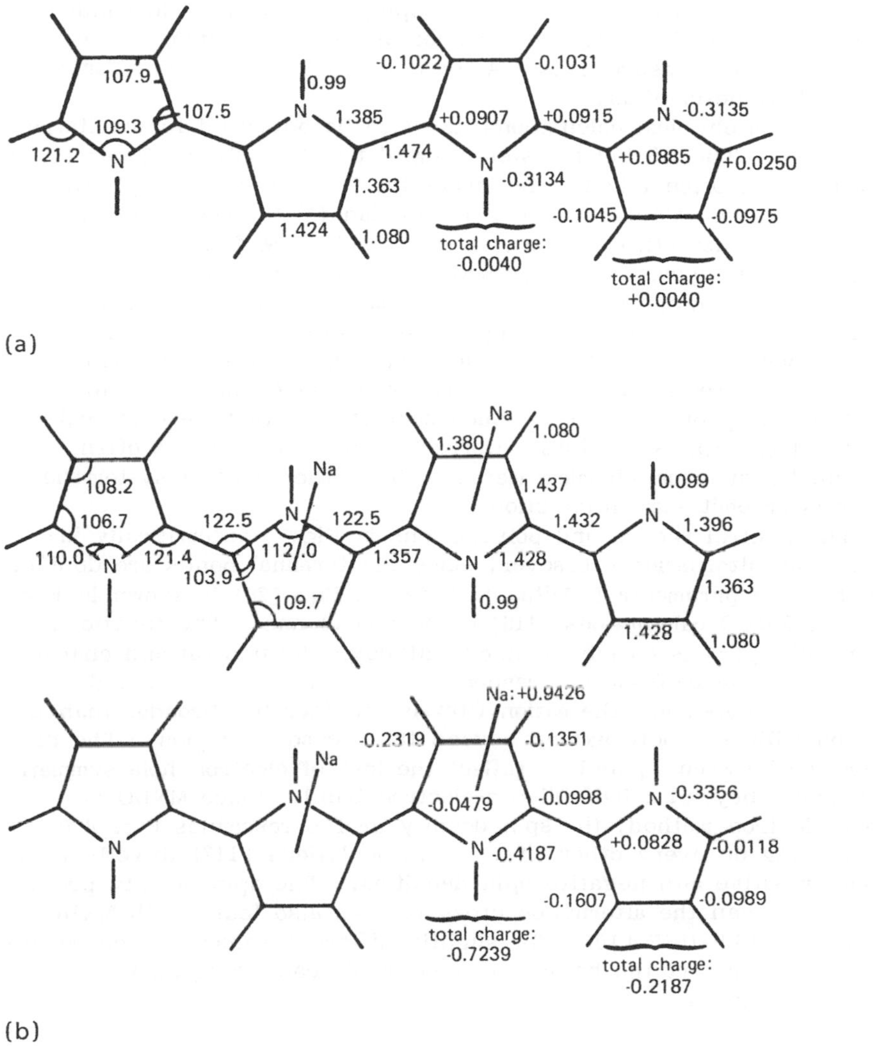

(a)

(b)

Fig. 16   (a) Optimized MNDO structure for linear pyrrole tetramer;
(b) optimized MNDO structure and charge distribution on adding two
Na above the inner ring; this corresponds to the limiting doping of
one Na per two pyrrole in the polymer.   (From Ref. 113.)

Solid-state descriptions [114,5] of doping start with rigid bands and emphasize the collective behavior of charged solitons to rationalize transport measurements. A great deal of work remains for a consistent description.

Quantum-chemical calculations may also be optimized for excitation energies. The Hückel transfer integral t in Eq. (2) and Table 1, for example, is often taken to be larger for thermodynamic applications [10]. In the valence effective Hamiltonian (VEH) method [115], completely theoretical and almost SCF results are found rapidly without iterative cycles. VEH methods have been extensively applied by Brédas and coworkers [99] to the optical gaps, ionization potentials, and electron affinities of conjugated polymers. The values in Table 3 are based on calculated or experimental geometries and include a constant correction of 1.9 ev to the ionization potential due to the polarizability of the lattice. Such corrections are typical for calculated (gas-phase) values. Correct relative ordering is often found by quantum-chemical methods that either contain systematic errors or omit some interaction.

The neutral radical and polyene ions in Fig. 5 automatically include an alternation crossover, since the terminal bonds are double. The order parameter $(-1)^p u_p \sim \Delta_S(x)$ in Eq. (20) is shown in Fig. 17 for MNDO calculations [116] on N = 41 chains. The functional form of $\Delta_S(x)$ is clearly retained, although the neutral and charged solitons have half-widths, respectively, of $\xi_O = 3$, $\xi_+ = 5$ and $\xi_- = 3$. Moreover, the cation CDW is considerably broader than the anion CDW and both extend beyond the crossover region. The difference between ξ_+ and ξ_- reflect the loss of electron-hole symmetry on going beyond a half-filled π-electron band. Since MNDO is a one-electron method, the spin density profile resembles Fig. 8 in vanishing on every other site. X-α calculations [117] directly give both positive and negative spin densities. The spin density profile is wider than the alternation crossover, as also found with MNDO [16] and PPP-UHF [103], for slightly different values. Such results strongly support the basic SSH ideas while cautioning against their literal acceptance.

C. Parameter Choices for Quantum Cell Models

The remarks above illustrate the central role of PA among conjugated polymers and of linear and cyclic polyenes among conjugated molecules. The chemical and topological simplicity of one $2p_z$ π-orbital per CH unit greatly reduces the number of parameters in Table 1, the values of which we consider below. Heteroatoms and large unit cells could be incorporated into quantum cell models, but such complications detract from the fundamental ideas of the SSH or continuum models discussed in Section II. PA and PDAs are the

Table 3   VEH results for the Ionization Potential, Width of the Highest Occupied Band, and Optical Gap of Hydrocarbon Polymers

Polymer	IP[a] (ev)	Bandwidth (ev)	E_g (ev)
Polyacetylene			
Trans	4.7 (4.7)[b]	6.5	1.4 (1.8)
Cis-transoid	4.8	6.4	1.5 (2.0)
Trans-cisoid	4.7	6.5	1.3
Polymethylacetylene	4.5	3.7	1.4
Polydiacetylene			
Acetylenic	5.1 (5.2)	3.9	2.1 (2.1)
Butatrienic	4.3	4.5	
Poly(p-phenylene)			
Coplanar	5.5	3.9	3.2
Twisted (22°)	5.6 (5.5)	3.5	3.5 (3.4)
Perpendicular	6.9	0.2	
Poly(m-phenylene)			
Coplanar	6.2	0.7	4.5
Twisted (28°)	6.2	0.2	(4.9)
Poly(p-phenylene vinylene)	5.1	2.8	2.5 (~3)
Poly(p-phenylene xylylidene	5.6	2.5	3.4
Polybenzyl	6.5	0.6	
Polyacene			
Regular	3.9	5.9	0
Trans	4.1	5.6	0.3
Cis	3.9	5.8	0
Polybiphenylene	3.4	5.9	0.1

[a]Theoretical ionization potential after subtracting 1.9 ev to correct, approximately, for polarization energy of lattice.
[b]Experimental estimates of IP and E_g are given in parentheses.
Source: Ref. 99.

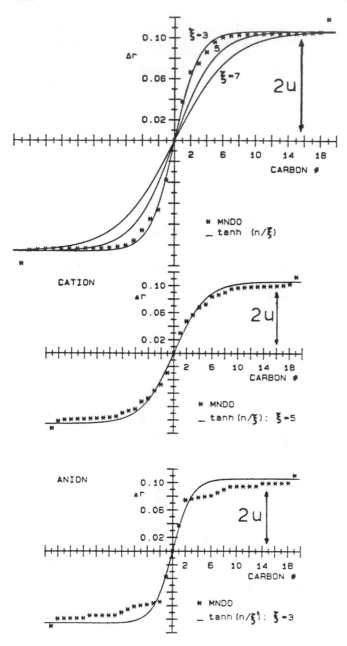

Fig. 17   MNDO alternation crossovers for neutral and charged solitons in N = 41 atom *trans*-PA chains.   The solid line from Eq. (23) give the crossover in the continuum model.   (From Ref. 116.)

simplest realizations of complicated systems. While linear and cyclic polyenes have provided initial tests for theories of electronic structure, many conjugated molecules offer comparable computational challenges and the restriction to π-electron models is no longer essential.

Thus π-electron models of PA as an infinite polyene are simultaneously the simplest problem in one field and the most difficult in another. The parameterization in Table 1 of the SSH model in Section II.B is phenomenological; the optical gap E_g and alternation u are combined with reasonable estimates and the bandwidth 4t and harmonic force constant K in Eq. (17). The Vanderbilt-Mele [118] parameters in Table 1 are chosen to fit also an optical phonon associated with the alternating SSH ground state, as summarized in Section III.E in connection with other consequences of e-p coupling. The emphasis is on an internally consistent model parameterized for trans-PA. As discussed by Wu and Kivelson [119], various SCF corrections for e-e interactions lead to additional parameters that actually exceed the available data points. Some independent estimates of parameters are soon needed in phenomenological models. Otherwise, different parameter choices and generalizations apply to other polymers, even for cis-PA with its nondegenerate ground state.

The PPP/molecular choices in Table 1, on the other hand, emphasize common parameter values for arbitrary π-electron systems with given ground-state geometry. Too many variations have been proposed over the years for a complete listing. A nontrivial test of π-electron models, even at the PPP level, is to achieve reasonable accuracy for conjugated hydrocarbons with t, α, V(R) in Eq. (9) fixed by the molecular geometry. Such consensus PPP values in Table 1 lead to polyene excitations in Table 2 and to the optical and magnetic data on low-lying states of naphthalene [84] in Table 4. The molecular parameters in Table 1 are also supported by direct calculations.

PPP models are consequently not expected to agree exactly with data on particular molecules, even when solved exactly. The triplet and singlet states of naphthalene in Table 4 could certainly be optimized by varying t, α, or V(R), but such variations would have to be understood in terms of special features of naphthalene compared to other π-systems. The very weak $^1B_{3u}$ oscillator strength [120] in Table 4 is unexpected at the one-electron level. It is associated with breaking the electron-hole symmetry by taking $e_9 = e_{10} = -0.15$ ev in Eq. (1) and leaving $e_p = 0$ for other p. Transition intensities are particularly sensitive to substituent (inductive) effects [121] that alter e_p. The observed [122,10] order of $E(^1B_{3u})$ and $E(^1B_{2u})$ is reversed in one-electron treatments. The actual ordering again illustrates degenerate $\varepsilon_1 + \varepsilon_2$ excitation falling below a nondegenerate ε_1 excitation due to e-e interactions as discussed in Section II.D for polyenes and radicals. The $^3B_{2u}$ fine-structure constants and spin densities in Table 4 were among the first EPR

Table 4   Comparison of PPP and Experimental Excitation Energies
ΔE, Magnetic Parameters, and Oscillator Strengths f of Low-Lying
Naphthalene States

State	Property	PPP	Experimental	Reference
$^3B_{2u}$	ΔE (ev)	2.522	2.64	10
Fine structure	D (cm^{-1})	0.1140	0.1012	123
	E (cm^{-1})	−0.0063	−0.0141	123
Spin density	ρ_1	0.4060	0.438	123, 63
	ρ_2	0.1164	0.125	10
	ρ_9	−0.0451	−0.126	10
$^1B_{3u}$	ΔE (ev)	3.604	3.96	120
	f^a	0.00063	0.0005	120
$^1B_{2u}$	ΔE (ev)	4.463	4.45	121
	f	0.2129	0.18	121

^{a}For $e_9 = e_{10} = -0.15$ in Eq. (1).

From Ref. 84.

applications to solids [123]. The exact $^3B_{2u}$ function is accurately,
but not exactly, given in PPP theory as a linear combination of the
29,700 triplet VB diagrams [84]. The corresponding SCF results
[124] are poor due to the near-degeneracy of several triplet states.

   The comparisons in Table 4 illustrate that the proper inclusion of
e-e contributions in excited states is preferable to reparameterization
of the model. The original PPP parameters [10,14] in Table 1 are
vindicated long after their apparent failure in one-electron approximations. Configuration interaction (CI) is also needed in all-electron
CNDO calculations [125] for the proper ordering of naphthalene
excitations. Optical excitations of both PA [126] and PDAs [127]
have received careful attention [5–8] and conjugated polymers may
eventually reach molecular resolution. At the π-electron level, the
molecular successes of the PPP model clearly make it the best choice
for conjugated polymers.

   There are various indications that molecular PPP parameters over-
estimate correlations in *trans*-PA. But simply reducing U in Eq.
(9) by 10% is hardly worthwhile. The explicit inclusion of a static

dielectric constant ε in $V(R) \sim e^2/\varepsilon R$ at large R is more interesting. In conductors, shielding of the mobile charges leads to a $\exp(-\kappa R)$ cutoff that fundamentally alters e-e contributions. A shielded potential has been proposed [102] to explain the low-energy onset of photoconductivity in *trans*-PA. Photoconductivity [128] and charge transport [114] have been difficult to model, however, and the appropriate solid-state version of Eq. (9) remains open. Improved PA preparation [129] leads to an almost 100-fold increase in the conductivity, essentially matching that of copper, and requires reevaluation of current models.

Although Hubbard models [19] were introduced for strongly-shielded 3d electrons in transition metals, their mathematical convenience has led to many solid-state applications to insulators or semiconductors. The appropriate "effective" U_e can always be defined for the most important correlation effect. In half-filled PA bands, for example, E_g and soliton spin densities suggest that $1.0 < U_e/|t| < 2.0$ on adding on-site interactions Eq. (6) to the SSH model [130], Eq. (17). Still smaller U_e may be assigned to purely solitonic effects to model deviations from mid-gap transitions [131]. The lowest ionic states of half-filled PPP models correspond to adjacent C^+C^- sites with Coulomb energy $U-V(R_0-u) \sim U_e$ and thus resemble an extended Hubbard model with fixed relative values of U and V. Such "strong" molecular correlations nevertheless have $U_e < 2|t|$. The range and form of the potential are at issue, rather than its strength. The need to retain at least a nearest-neighbor V has become more widely appreciated in both conjugated polymers and π-molecular solids. Unless there is screening, short-range potentials are a matter of convenience.

D. Chain Length, Conformation, and Interchain Coupling

The SSH or continuum model for PA is based on an infinite extended chain, whose band structure is sketched in Fig. 4. Solid-state models of conjugated polymers almost always invoke a one-dimensional crystal and its bands. Polymeric strands are neither infinite nor extended in practice. We first consider the implications of long but finite segments, comment on interchain interactions, and then turn to conformational degrees of freedom associated with bond rotation. While periodic one-dimensional representations will undoubtedly remain basic to conjugated polymers, we anticipate greater care in applications to phenomena in noncrystalline samples.

The conjugation length of real polymers is inevitably limited by cross links, chain ends, and chemical or structural defects. The conjugation length in PA is about $N \sim 100-300$ CH units, while crystalline PDA samples may have $N > 10^3$. Finite-length effects are often invoked, for example to rationalize the blue shift [6,7] of E_g at the rod-to-coil transition of PDA solutions or films as a consequence of

shorter chain lengths, although the microscopic description of the
defects remains speculative. The order of magnitude of N, rather
than its exact value, is important here. Thus $N \geqslant 10^2$ is large
enough for infinite-chain optical gaps in Eq. (26), in Fig. 13 and
for the band structure in Fig. 4. Finite N is a nuisance in this
context.

Finite chain lengths [132,133] have serious implications for alternation crossovers. The magnetic susceptibility [134] of *trans*-PA
corresponds to about one unpaired electron per 3000 carbons. Since
adjacent odd segments (radicals) are expected to recombine by forming cross links, thereby generating even conjugation lengths between
sp^3 centers, most segments have even N. For $N \sim 10^2$, there are
some 30 even (e) segments per odd (o) one. The ground state of
even polyenes is not degenerate, however, since interchanging double
and single bonds in Fig. 1 produces a biradical with higher energy.
Degenerate ground states are restricted to finite even rings or to
$N \rightarrow \infty$. A pair of neutral solitons, such as the biradical formed on
interchanging double and single bonds in an even polyene, is a
short-lived excited state. The bond orders [135] of the 2^1A_g state
have this alternation pattern and its lifetime [136] can be measured
in *trans*-octatetraene.

The postulated mobility of alternation crossovers leads to rapid
(picosecond) geminate recombination [137,138] of electron-hole pairs
produced by photons $\hbar\omega \geqslant E_g$. Trapping is unavoidable in one-
dimensional systems. Pairs of neutral solitons on finite segments
must recombine to form the nondegenerate ground state. Except
immediately after excitation, *trans*-PA has no solitons associated
with even segments and a single alternation crossover confined to
each odd segment. Soliton motion on odd chains is probably free,
except for higher energy near the chain ends [60], for N large
compared to the soliton width ξ discussed in Section II.B. Nutation
^{13}C NMR studies [40] show no interconversion of double and single
bonds in *trans*-PA even at 300K. As emphasized by Clarke and
Scott [139], this excludes soliton motion in most ($\sim 90\%$) of the
sample. The restriction of mobile solitons to odd segments comprising less than 10% of the sample, on the other hand, cannot be ruled
out. Although attractive chemically, a segment model loses most of
the field-theoretical connections of Section II.C and becomes a
problem in static defects.

The occurrence of even and odd segments of $N \geqslant 10^2$ carbons also
bears on PA doping. Campbell and coworkers [140] have emphasized
that *polarons* are initially produced in PA on doping with either electron donors or acceptors, which certainly holds for even segments
or for infinite chains. The original SSH proposal [13] was for
dopants to convert neutral S to S^+ or S^- by removing or adding an
electron to the midgap state in Fig. 6. A concentration [o] of fixed
odd segments is preferentially converted on doping to o^+ or o^- even

on including e-e correlations [132]. The electron-hole symmetry of linear PPP or extended Hubbard models leads to

$$I_e = I_o + \Delta(U/4|t|)$$

$$A_e = A_o - \Delta(U/4|t|)$$

$$(27)$$

for the ionization potential I and electron affinity A of even and odd chains with $N \to \infty$. We have $\Delta(0) = E_g/2 \sim 0.9$ ev in the band limit and $\Delta(\infty) = 0$ in the atomic limit. The value of Δ is intermediate for standard molecular PPP parameters from Table 1. Neutral odd segments thus scavenge [132] charge by being the first to accept or lose electrons from dopants, as sketched in Fig. 18. The more numerous even segments are ionized only at higher doping. The paramagnetism of *trans*-PA implies the existence of neutral odd segments and the initial formation of S^+ or S^- as o^+ or o^- ground states, respectively.

The pinning potential V_i ($\leqslant 0.5$ ev) between S^+ or S^- and the dopant ion does not change the initial formation of o^+ or o^- segments. The delocalized soliton charge over some $2\xi \sim 15$ sites reduces V_i and the dependence on the impurity-soliton separation R_{si}. For approximately one odd chain in 15 (N = 200) or 30 (N = 100) and chain coordination z = 6, most dopants are adjacent or second-neighbor to an odd segment and are automatically pinned. The

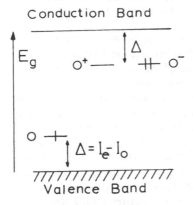

Fig. 18   Schematic PA band structure for a low concentration of odd segments among a majority of even segments; the higher electron affinity of o^- and lower ionization potential of o are given by Eq. (27). Trap-to-trap hopping of o^+ or o^- to o is a mechanism for charge transport among noninteracting segments.

competition between V_i and Δ in Eq. (27) determines whether an unfavorably located odd segment is ionized instead of an even segment. At finite temperature, the entropic consequences of more even than odd segments and a distribution of V_i pose statistical problems.

The emerging picture for *trans*-PA is a mixture of mostly even and a few odd polyenes, all long enough (N $\gtrsim$ 10^2) to ignore simple length effects but fundamentally different in having singlet and doublet ground states. The far smaller paramagnetism [134] of pristine *cis*-PA points to fewer odd segments. Polymerization mechanisms [141] imply C_2H_2 addition and an initially cis geometry. In the approximation of noninteracting strands, *trans*-PA becomes a mixed molecular solid whose optical excitations are the superposition of neutral even and odd segments with comparable E_g, and also of o^+ or o^- if lightly doped. Adventitious p-doping converts some 5–10% of odd segments to o^+, whose low-energy absorption was discussed in Section II.D and disappears on compensating with NH_3.

Many recent studies [142,143] on *trans*-PA and crystalline PDAs have dealt with the position and dynamics of photogenerated states. Either points of agreement or disagreement with specific models can readily be cited. Friend and coworkers [144] have confirmed that interchain electron transfer are important processes. This can readily be added to most single-strand models. Not so easily treated quantitatively, on the other hand, is the altered density of states or new mechanisms for interchain charge transport. Interchain hopping of bipolarons may be favored [145] over charged solitons, for example, since the geometrical changes (Franck-Condon factors) for bipolarons in Fig. 9c are restricted to the region between the two alternation reversals.

Since odd segments have lower I and higher A in Eq. (27) than even segments, charge transport at low dopant concentration can be formulated in Fig. 18 as transfers between o^+ or o^- and nonneighbor o segments. Such three-dimensional transport via deep traps has been studied for singlet or triplet excitons in mixed molecular cystals [146], often isotope-labeled pairs whose orientation does not disrupt the lattice and produces shifts through differences of zero-point vibrational energies. Emin [147] has discussed hopping models for charge transport in *trans*-PA without identifying the localized states or traps. Other phenomenological approaches [1–5] to transport and doping are based on charged solitons, polarons, or bipolarons in liquid or lattice states. They rationalize some features and are readily changed as new observations are made. Interchain interactions, intersegment contacts, and polymer morphology are all relevant.

We turn next to conformational degrees of freedom and their effect on the optical gap. Rotations about nominally single bonds in PA or PDA connect isomers with almost equal (<0.1 ev) energies [148,149] and involve modest barriers. Except in single crystals, the extended geometries implied in Figs. 1, 2, or 14 are convenient fictions. The conformations of nonconjugated polymers have long been discussed [150] either in terms of continuous semiflexible strands, as sketched in Fig. 19, or in terms of specific bond rotations and excluded volume reflecting the chemical nature of the polymer. Flexible or wormlike polymers are defined through the autocorrelation function

$$\langle \hat{u}(s) \cdot \hat{u}(s') \rangle_c \; = \; \exp(-2\lambda \, |s-s'|) \tag{28}$$

The vector $\hat{u}(s) = d\vec{r}(s)/ds$ is the unit tangent to the polymer backbone at s, $(2\lambda)^{-1}$ is the persistence length and is longer than R_0, $\langle \cdots \rangle_c$ is the configurational average, and $0 \leqslant s \leqslant L$ corresponds to the polymer length of $(N - 1)R_0$ for N monomers and spacing R_0. The rod limit $(\lambda \to 0)$ corresponds to extended segments or one-dimensional crystals whose characteristic band structure is sketched in Fig. 4. *All* considerations of optical excitations in Section II are based on $\lambda \to 0$. The resulting selection rule for crystals is to vertical transitions, or $\Delta k = 0$, in Fig. 4. We may in principle invoke a two-dimensional representation in Fig. 19 in which there is no change of conjugation, only a distribution of cis and trans double bonds. But a three-dimensional picture of the polymer is more realistic.

The consequences of finite λ in alternating Hückel models for conjugated polymers have recently been discussed by Soos and Schweizer [151]. Bond rotations in Hückel, Hubbard, or extended Hubbard models leave the potential $V(R)$ unchanged and thus cannot alter the overall spectrum in Fig. 4, which satisfies a sum rule. More distant interaction $V(R)$ in neutral PPP models lead to small energy changes that are neglected. The crystal selection rule is relaxed to $\Delta k \leqslant \lambda$, thereby formally including nonvertical excitations

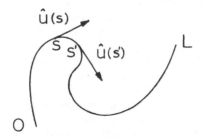

Fig. 19 Schematic representation of a flexible polymer. The autocorrelation function, Eq. (28) is the unit tangent to the backbone and defines the persistence length.

in Fig. 4. The greater phase space for $\Delta k \leqslant \lambda$ shifts the average absorption to higher energy. The rod ($\lambda = 0$) transition at $E_{\vec{g}}$ is blue-shifted with increasing λ due to purely conformational contributions to the dipole transition moment. Physically, the $\pi \to \pi^*$ absorption intensity reflects the relative orientation of the oscillating electric field, $\vec{E}(\omega)$, and the three-dimensional structure of the polymer.

Rotational defects in conjugated polymers may in principle completely disrupt π-conjugation, with $t_p \backsim 0$ at single bonds rotated by $\pi/2$ in Fig. 1 or 2. Kuzmany [152] and Mulazzi et al. [153] interpret resonance Raman data on *trans*-PA in terms of a bimodal distribution of short (N $\backsim$ 10) and long (N $\backsim$ 10^2) segments. Segment models have also been applied to the rod-to-coil transition of several PDAs [154] and σ-conjugated polysilane [155] systems. Such strong disorder is convenient mathematically rather than realistic physically. The chemical lengths of PA strands may, for example, be known [156] to be the same in stretched and amorphous samples.

The coupling of electronic structure and polymer conformation is characteristic of conjugated polymers and reflects the sensitivity of π-overlap to bond rotation. Since single bonds have low barriers to rotation, increasing λ in Eq. (28) selectively reduces $|t_p|$ at single bonds and thus increases [151] the effective alternation as $\delta(\lambda) = \delta + 2A\lambda$, where A is a constant around 0.5. Within a flexible strand approach, the polymer conformation blue-shifts the $\pi \to \pi^*$ absorption both by relaxing the selection rule to $\Delta k \leqslant \lambda$ and by increasing the effective alternation. While e-p and e-e contributions remain to be included, realistic descriptions of the optical excitations of conjugated polymers must explicitly include conformational degrees of freedom.

E. Enhanced Vibrational Excitations

We discussed bond-length alternation of polyenes in Section II.A in terms of π-electrons in a half-filled band. The σ-electron contribution for the SSH model in Eq. (17) is harmonic by hypothesis and is minimized for x = 0, or uniform bond lengths R_0. The ground-state energy per site is doubly degenerate, as sketched in Fig. 20, with alternating bond lengths $R_0 \pm u$ along the infinite chain. Since only the distance dependence of t(R) is kept in Eq. (3), it is natural to restrict the vibrational problem to carbon-carbon stretches in a linear chain, as suggested in Fig. 20. Vibrations about the minima at $\pm u$ correspond to b_u phonons that are strongly coupled to the π-system. Thus e-p coupling emphasized in the SSH and continuum models has important vibrational consequences. Both the

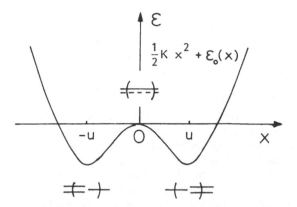

Fig. 20   Schematic representation of the ground-state potential of PA for alternation x.

polarizability of the π-system and the symmetry breaking associated with alternation crossovers may be probed by infrared and Raman spectroscopy.

There are many aspects to vibrational studies. One has been to model the multidimensional ground-state potential surface, as reviewed for short polyenes by Hudson et al. [83]. The all-trans even polyene $C_{2n}H_{2n+2}$ in Fig. 1, for example, has vibrational normal modes given by

$$\Gamma = (4n + 1)a_g + 2na_u + (2n - 1)b_g + 4nb_u \qquad (29)$$

The in-plane a_g modes and out-of-plane b_g modes are Raman-active, while the in-plane b_u and out-of-plane a_u are IR-active. Most of these modes have been observed and modeled for butadiene (n = 2), *cis*- and *trans*-hexatriene (n = 3), and *trans*-octatetraene (n = 4). Even more complete vibrational analysis is available for the benzene ground state [157]. The force constant K in Table 1 for b_u vibrations representing alternation fluctuations in benzene can consequently be taken from experiment. The systematic modeling of the carbon-carbon stretches with increasing length remains a goal [83]. Force constants for bond stretching and bending must be included, as well as various second-neighbor contributions.

Two points are worth noting in the present context: (a) the PA potential in Fig. 20 is highly idealized, and (b) the force constants for molecular b_u vibrations are known. Vibronic coupling in molecular excited states is another important topic that is beyond the scope of the present review. The elementary excitations of the continuum limit in Section II.C lead to local distortions $\Delta_s(x)$ in Eq. (23) for solitons and to $\Delta_p(x)$ in Eq. (24) for polarons, again in terms of a one-dimensional model restricted to carbon-carbon stretches. Although the C–C and C≡C stretches are most sensitive to electronic excitation, many other octatetraene vibrations in the 2^1A_g and 1^1B_u excited states are shifted slightly [80].

We summarize two related approaches for the idealized PA ground-state potential in Fig. 20. Mele and Rice [158] consider noninteracting π-electrons (SSH model) and emphasize the π-electron polarizability contribution to the dynamic matrix. They consider five force constants, including C–H stretches and bends, taken from small molecules. Phonon spectra are changed most near the band center (q = 0), where the frequency of the optical (b_u) mode is strongly depressed. The curvature at ±u in Fig. 20 then gives the frequency

$$\Omega(0) = (16\alpha^2/\pi tM)^{1/2} \tag{30}$$

for the b_u mode in terms of the CH mass M, linear e-p coupling constant α, and transfer integral $t(R_0)$. Quite generally, a quantitative model for the PA bond-length alternation also predicts the b_u frequency. Vanderbilt and Mele [118] show that the original SSH parameters give a poor fit to $\Omega(0)$ and propose revised value given in Table 1. Their K = 68.6 ev/A^2 is unfortunately far larger than measured in small molecules. Heeger [95] also favors larger e-p coupling $\alpha \sim 8$ ev/A for understanding the IR. There are again limits to quantitative applications of the SSH model, at least if similar microscopic parameters are accepted in conjugated molecules and polymers.

Takahashi and Paldus [159], on the other hand, start with interacting π-electrons and adopt molecular parameters in the PPP model. The π-electron ground-state energy per site $\varepsilon_0(x)$ in Fig. 20 is evaluated for 4n+2-site rings using various [101] RHF, UHF, and alternant MO schemes. An anharmonic σ-potential is chosen to reproduce the hexagonal (undimerized) benzene geometry and b_u vibration. Longer 4n+2 rings eventually show a Peierls instability and have double-well potentials as sketched in Fig. 20. The ground-state alternation u $\sim$ 0.05A converges by n $\sim$ 10–12 and the infrared frequencies are reasonable for both cis- and trans-PA. The e-p coupling constant α is effectively enhanced by both self-polarizability and potential contributions. The small PPP/molecular value in Table 1 does as well as the Vanderbilt-Mele choice for noninteracting electrons. The important point is that molecular parameters suffice for the PA equilibrium and infrared data. As shown in Section II.D, such parameters also account for low-lying electronic states when e-e correlations are fully included. Undoped PA and other conjugated polymers are giant π-electron molecules as expected chemically. Widespread transferability of microscopic parameters has obvious advantages in guiding experiment and in testing theoretical models.

Mele [158] has reviewed other applications to study localized phonon modes associated with a soliton in PA. An important result is that the soliton breaks the inversion symmetry of the chain. The same vibration may then be seen in both IR and Raman spectra. Charge fluctuations along the backbone strongly enhance a_g modes in π-radical solids also [160]. Changes in IR or Raman spectra on doping or photoexcitation thus directly probe charged excitations even in the absence of a complete vibrational analysis.

Horowitz and coworkers [161,162] have exploited a phenomenological description of a_g phonons to focus explicitly on changes due to charged excitations. The amplitude mode $\Delta(x)$ in Eq. (20) is directly linked to the alternation in the continuum model. Vibrations about $\pm u$ in Fig. 20 correspond to a $q = 0$ modulation of the optical gap $E_g = 4t\delta = 4\alpha u$. Horowitz [161] supposes p phonon fields to be sensitive to such modulation and finds $p = 3$ in *trans*-PA. Batchelder [163] reports $p = 4$ Raman-active modes for the PDA backbone. By expanding about the ground-state (dimerized) solution of the continuum equations, Horowitz finds the frequencies of $q = 0$ phonons in the adiabatic limit. As shown in Fig. 21, the three Raman-active lines are sensitive to the frequency $\hbar\omega = E_g$ of the exciting laser. The shifts are solutions of

$$D_0(\omega) = \Sigma \frac{\lambda_p}{\lambda} \frac{(\omega_p^0)^2}{\omega^2 - (\omega_p^0)^2} = \frac{-1}{1 - 2\tilde{\lambda}} \tag{31}$$

where ω_p^0 and λ_p/λ are, respectively, adjustable parameters in Table 5 for the phonon frequencies and dimensionless coupling constants. The electron-phonon coupling constant $\tilde{\lambda}$ is given by

$$1 - 2\tilde{\lambda} = 2\lambda\varepsilon_0''(u) \tag{32}$$

and depends on the curvature of the electronic ground-state energy per site. For noninteracting electrons, $\tilde{\lambda}$ reduces to λ and ε_0 is given by Eq. (15). The excellent fits in Fig. 21 are extended in Table 5 to IR modes. The pinning constant α is the only new parameter, with $D_0(\omega) = 1/(\alpha - 1)$ in Eq. (31). Physically, charged soliton are confined to be near dopants, as mentioned in Section III.D, and pinning is reduced for photogenerated states.

Such a phenomenological approach focuses on the fewest parameters needed for a large data set. The microscopic origin of α and the

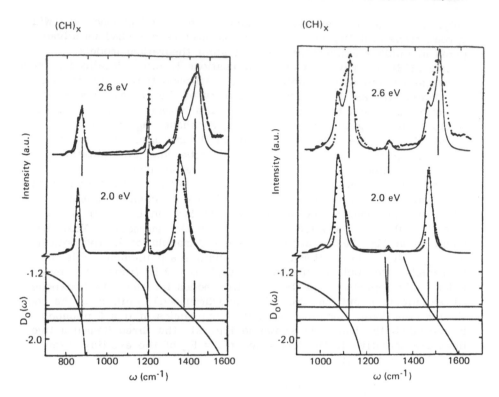

Fig. 21   Resonance Raman spectra and fits based on Eq. (31) for $(CH)_x$ and $(CD)_x$. (From Ref. 161.)

molecular vibrations corresponding to ω_p^0 are separate problems.

The fits in Fig. 21, in Table 5, and in subsequent [164] refinements and extensions to include disorder certainly invite such derivations. Horowitz has also pointed out that the dependence in Fig. 21 on the Raman frequency $\hbar\omega$ is consistent with $\tilde{\lambda} = \lambda$ in Eq. (32), and the assumption of a sharp, δ-function absorption at $\hbar\omega = E_g$. The latter is an oversimplification. The optical gaps of some chains, are then well above E_g, presumably due to some combination of finite-chain, configurational, or environmental effects.

The absence of e-e contributions in such quantitative applications is interesting. To be sure, $\varepsilon_0''(u)$ in Eq. (32) involves a ground-state property and is thus less sensitive to e-e correlations. The closed electronic shells of most molecules facilitates reliable quantum chemical results for bond lengths and angles in Section II.B. The Hückel result for $\varepsilon_0(u\alpha/t)$ in Eq. (15) leads immediately to the

Table 5 Induced IR-Active Phonon Frequencies (in cm^{-1}) by Doping or Photogeneration

Parameters (in Eq. 31)

ω_n^0	λ_n/λ	IR-doping $\alpha = 0.23$		IR-photogeneration $\alpha = 0.06$	
		Theory	Experiment	Theory	Experiment
(CH)$_x$					
1234	0.07	886	888	488	~500
1309	0.02	1285	1288	1278	1275
2040	0.91	1397	1397	1364	1365
(CD)$_x$					
921	0.06	770	790	410	~400
1207	0.005	1148	1140	1045	1045
2040	0.93	1236	1240	1216	1224

Source: Ref. 161.

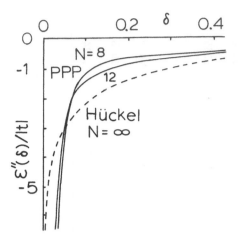

Fig. 22   Curvature of the electronic ground-state energy per site as a function of alternation for Hückel chains, Eq. (15) and for N = 4n PPP rings with molecular parameters.

curvature in Fig. 22. The logarithmic divergence as u → 0 ensures an alternating ground state. The molecular PPP potential for a regular array and variable t(1 ± δ) leads [165] to similar curvatures in Fig. 22 around δ = 0.10 for rings of N = 8 and 12. The optical gap of 4n rings is independent of n in Hückel theory, as can be seen from $k_F = \pi/2$ in Fig. 4, and directly reflects e-e contributions. Extrapolation of the PPP curves in Fig. 22 moves them even closer to the Hückel curve. Even without extrapolation, however, their similar behavior around δ ∿ 0.1 shows ε'(u) and thus $\tilde{\lambda}$ in Eq. (32) to be insensitive to correlations.

F. Low-Energy PDA States

Wegner and coworkers [166] developed the solid-state polymerization of PDAs and found suitable R-groups in Fig. 2 for forming single crystals. The advantages of crystalline samples were soon recognized by Bloor and coworkers [127,163] for optical and vibrational studies and by Wilson and coworkers for photoconduction. Suggestions of high-energy butatriene (BUT) polymers in Fig. 11b had to be revised, as the acetylenic PDA structure was found [167] to be characteristic of crystals. Chemical doping of PDAs has been difficult [168] and precludes parallel studies of insulator-to-metal transitions typical of conjugated polymers. Strong parallels to PA are found in the behavior of photogenerated states [142,143] and in nonlinear electrical properties [5]. The simplicity of the PDA backbone invites direct application of quantum cell models, in contrast to the chemically and structurally more complicated polymers in Fig. 14. The joint analysis of PA and PDA puts current models and understanding into perspective.

The triple bond of C_2H_2 is about 1.20 A, and such short triple bonds occur [168] in PDAs. The triple bond clearly increases the alternation and bandwidth W compared to *cis*- or *trans*-PA, even though linear e-p coupling in Eq. (3) may seriously underestimate the changes in transfer integrals. Since e-e contributions appear to be similar to those in PA, PDAs are less correlated and molecular PPP parameters now place [169] 2^1A_g above 1^1B_u (or E_g) rather than in the polyene order discussed in Section II.D. Different experimental methods [170], all indirect, place a state thought to be 2^1A_g on either side of the optical gap. The position of 2^1A_g is not known accurately in PA either, although it is generally assigned [71] below E_g.

Increased alternation increases E_g in either Hückel or PPP models. But the strong PDA $\pi \to \pi^*$ absorption [127] to 1^1B_u is at 2.0 ev, just above the *trans*-PA peak at 1.9 ev and just below the *cis*-PA maximum around 2.1 ev. The same t(R) curve in Eq. (3) cannot account for both PA and PDA. Moreover, the PDA excitation is excitonic [172], with a bound electron-hole pair, since the onset of photoconduction is a few tenths of an ev higher. The *trans*-PA threshold [173] for photoconduction is well below E_g, which has been interpreted as a band-to-band transition [138]. Models without e-e interactions cannot account for excitons. The different nature of the $\pi \to \pi^*$ excitation, the photoconductivity threshold, and the absence of effects due to increased alternation, on the other hand, raise questions about quantum cell models of similar conjugated systems.

The triplet state 1^3B_u may be produced in PDA and studied [174, 175] via optically detected EPR, with a singlet-triplet gap of about 1.1 ev. It is also the lowest electronic state of polyenes, at ~ 1.9 ev for *trans*-octatetraene [83], for naphthalene in Table 4, and indeed generally for π-systems with singlet ground states. The lowest triplet state of longer polyenes or of PA has yet to be detected. Such correlation effects are included in PPP models, may readily be rationalized by perturbation theory, and are essentially Hund's rule.

The photopolymerization of diacetylene crystals can be controlled and quenched at low temperature. Sixl [75] used EPR and optical techniques to obtain detailed information about the formation of diacetylene dimers, trimers, etc. At low temperature, the diradicals DR_m in Fig. 23 with m = 2, 3, 4, and 5 are stable. Their BUT structure is consistent with the fine-structure of the terminal unpaired electron and with low-energy optical excitations; the singlet and triplet states of DR_m are almost degenerate. Longer fragments

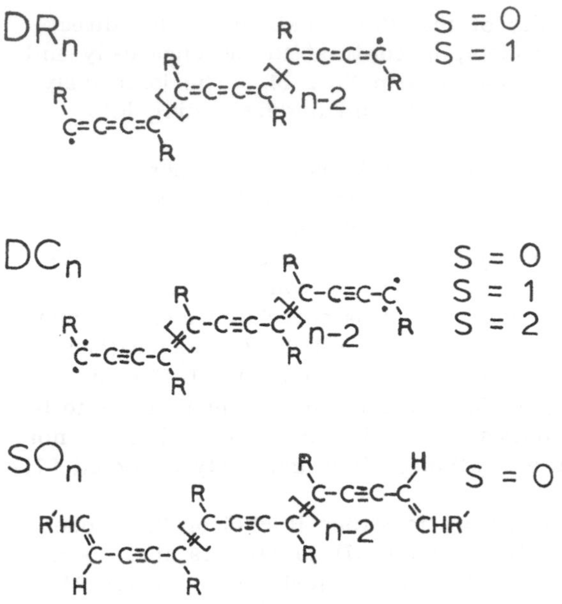

Fig. 23   Schematic representation of diradicals (DR$_n$), dicarbenes (DC$_n$) and single toligomers (SO$_n$) formed during the photopolymerization of diacetylenes.   (From Ref. 75.)

than m = 6, on the other hand, form dicarbenes DC$_m$ in Fig. 23, with fine-structure constants indicative of two unpaired electrons on the terminal carbons.   The lower-energy PDA form evidently compensates the extra π-bond in DR$_m$ at m $\simeq$ 6.   An H-atom shift from the α-carbon of R is supported by a strong isotope effect and leads to stable oligomers SO$_m$ in Fig. 23 with N = 4m + 2 conjugated carbons.   The lowest $\pi \rightarrow \pi^*$ excitation of SO$_m$ with m = 2, 3, and 4, are, respectively, at 3.0, 2.7, and 2.5 ev.   They fall *below* the lowest excitation of the corresponding polyene in Table 2, even after adjusting the latter to solids.   In solution, the lowest $\pi \rightarrow \pi^*$ absorptions are broader and quite similar for polyenes and ene-ynes with the same conjugation length [176,177].

Thus quantum cell models with a single ϕ_p per site for PDA cannot share the microscopic parameters of conjugated systems in Table 1.   Alternatively, common parameters require a more general model. One possibility is to retain [169] a ϕ_p for each π-electron at triple bonds in Fig. 2.   The extended π-system has a $2p_z$ orbital at each site normal to the PDA plane, while π'-electrons occur in separate pairs at triple bonds and merely contribute to the core electrons in an SCF approximation.   Although π'-electrons are assigned formally to

σ-orbitals in nonlinear molecules, the identity of the π' orbitals is well preserved in small ene-ynes [178]. Polymerization of diacetylenes leads to a new high-energy excitation around 7 ev whose polarization is along the polymer backbone [179]. A π'-dimer with R = 1.20 and standard PPP parameters [169] absorbs at 7.8 ev.

Electron transfers t' between the orthogonal π and π'-system vanish in the first approximation. The Coulomb potential V(R) in Eq. (9) couples the two subsystems. A single transfer in C $\equiv$ C leads to a higher-energy $C^+ = C^-$ diagram; transfer in the other π-systems give a formally neutral dicarbene $\ddot{C}-\ddot{C}$ or a very high-energy $C^{+2}-C^{-2}$ diagram. Dicarbene (pseudocovalent) VB diagrams are strongly admixed in such extended PPP models [169]. They lower the energy at dipole-allowed 1^1B_u and 2^1B_u states, make their intensities more nearly equal, and shift the lowest $\pi \to \pi^*$ absorption of short segments below that of the corresponding polyene. The 2^1A_g and 1^3B_u states are shifted to higher energy by $\pi - \pi'$ coupling [180]. Coulomb coupling between the extended π'-dimers account well for the lowest ene-yne excitations and invite applications to PDA.

The contrast between the low-energy PDA and PA excitations illustrates both good and bad features of π-electron models. The qualitative consequences of alternation crossovers in polymers with non-degenerate ground states rationalize several aspects of the photo-generated states. But the expected similarities, expecially at the Hückel or SCF level when the π'-electrons are well below ε_F, are missing. Different parameter choices raise questions about their interpretation in chemically similar cases. Alternatively, general considerations must justify an expanded basis. The choice of quantum cell models and of their parameters both require insight. Successful models presuppose considerable understanding and experience with the physical system.

IV. CONCLUDING REMARKS

We noted at the outset the many guises of π-electron models of conjugated polymers in widely different contexts. We have introduced and contrasted various points of view, especially those associated loosely with chemists and physicists, of low-lying electronic states. Detailed experimental or theoretical comparisons have deliberately been avoided, even on topics explicitly covered, and left to the references. We regret the inevitable incompleteness. In addition, entire areas have been virtually omitted. These include doping, insulator-to-metal transitions, batteries and other electrochemical applications, polymer morphology and phase changes, the polyaniline family, transient and steady-state transport measurements and their interpretation, and many nonlinear electrical responses. To some degree, they involve the relaxation or dynamics of low-lying

π-excitations, interfaces between phases, or higher-energy excitations.
We have concentrated on the position and nature of the lowest
π-electron states. Static properties offer simpler and sharper applications of quantum cell models. Moreover, many kinetic schemes for
transport are based on assumed diffusion, motion, collisions, and
lifetimes for such excitations.

The central importance of alternation crossovers or pairs of crossovers for excess charge in conjugated polymers is clear from quantum
chemical calculations as well as from SSH and continuum models.
Valence bond diagrams associated with charged solitons, polarons, or
bipolarons are also immediately recognizable but differently named by
organic chemists and may almost be traced to Pauling some 50 years
ago. Valence bond pictures of alternation crossovers thus provide a
common language for discussing transport and excitations in conjugated networks. Such quantitative aspects of crossovers as their
width, geometry, wavefunction, or energy fortunately need not be
specified too accurately in current applications.

Even at the π-electron level, both electron-phonon and electron-
electron contributions are important. We have sketched qualitative
changes due to each. It may be true that e-e contributions are
less important for most ground-state properties, or that e-p effects
in excited states can largely be incorporated by readjusting some
transfer integrals. Such *a posteriori* arguments merely demonstrate
that some simplified, usually single-particle, description can be con-
cocted once the collective properties are correctly understood. The
revised model parameters then contain the interesting physics. The
joint analysis of e-p and e-e interactions on as equal footing as
possible is equally appealing and equally difficult for both physicists
and chemists. New mathematical models are needed. Some recent
computational advances include various Monte-Carlo methods, exact
solutions of finite systems, renormalization group methods [181], and
interpolation schemes [182]. Such methods are hardly brute force,
but pay careful attention to exact solutions and theorems about
quantum cell models.

The proper role of quantum cell models *vis-a-vis* experimental observation is beyond resolution and involves taste and judgment. Any
π-electron model is inherently approximate, although it may be excellent for the lowest excitations. The proper choice of ϕ_p is not obvious in PDA, for example. The parameters may be fine-tuned to
an important special case like *trans*-PA or related to the molecular
geometry of arbitrary conjugated molecules. Even then, exact solutions of quantum cell models hardly exist for conjugated polymers
and additional approximations are necessary. So the model, its
parameters, or its solution must all be considered for experimental
comparisons, while only approximations must be considered when comparing to other calculations. Yet experimental observations and

understanding of conjugated polymers are the main reason for constructing models.

Theoretical analyses of idealized single strands, including polyacenes [183] and other conjugated ribbon polymers [184] whose preparation is far from assured, will continue to be important. The Peierls instability is greatly reduced in Hückel models of polyacenes due to the nodal properties of the HOMO and LUMO. Extended Hückel band structures may in principle identify systems with small E_g and such tuning has been discussed for derivatives of polythiophene. Theorists are properly reluctant to discard prematurely the convenience of periodic one-dimensional arrays. Features that are long compared to bond lengths, such as alternation crossovers or the persistence length of flexible chains, may sensibly be discussed in a continuum limit.

We anticipate greater appreciation of chain length and conformation in future work on conjugated polymers whose strands are neither infinite nor extended. Hückel models may suffice for the main features, which can then be refined by including e-e and e-p effects in quantum cell models. Interchain interactions are likely to remain perturbations to inherently one-dimensional models. The focus of quantum-chemical descriptions on local structure and field-theoretical considerations of continuum models should persist indefinitely, since different length scales are involved. Both provide insights about the electronic structure of conjugated polymers and both will eventually be part of the standard model.

ACKNOWLEDGMENTS

The support of the National Science Foundation (DMR-8403819) for work at Princeton is gratefully acknowledged. The development and application of diagrammatic VB methods to polyenes and to polymers owes much to Drs. Sumit Mazumdar, S. Ramashesha, Stephen Bondeson, Laurent Ducasse, and Satoru Kuwajima.

REFERENCES AND NOTES

1. T. A. Skotheim (ed.), *Handbook of Conducting Polymers*, Vols. 1 and 2, Marcel Dekker, New York, 1986.

2. H. J. Cantow (ed.), *Polydiacetylenes, Advances in Polymer Science*, Vol. 63, Springer-Verlag, Berlin, 1984.

3. R. H. Baughman, J. L. Brédas, R. R. Chance, R. L. Elsenbaumer, and L. W. Shacklette, *Chem. Rev. 82*:209 (1982).

4. (a) S. Etemad, A. J. Heeger, and A. G. MacDiarmid, *Ann. Rev. Phys. Chem. 33*: 433 (1982); (b) J. L. Brédas and G. B. Street,

Accts. Chem. Res. 18:309 (1985); (c) A. O. Patil, A. J.
Heeger, and F. Wudl, *Chem. Rev.* (to be published).

5. Proceedings of the International Conference on Science and
   Technology of Synthetic Metals, Kyoto, Japan (1986), *Syn. Met.
   17–19* (1987). The author and subject indices are particularly
   useful.

6. Proceedings of the International Conference on the Physics and
   Chemistry of Low-Dimensional Synthetic Metals, Abano Terme,
   Italy, *Mol. Cryst. Liq. Cryst., 117–121* (1985).

7. D. Bloor and R. Chance (eds.), *Polydiacetylenes*, NATO ASI
   Series E102, Martinus Najhoff, Dordrecht, The Netherlands,
   1985.

8. D. J. Sandman (ed.), *Crystallographically Ordered Polymers*,
   ACS Symposium Series 337, American Chemical Society, Washington, D.C., 1987.

9. D. Williams, ed.), *Nonlinear Optical Properties of Organic and
   Polymeric Materials*, Plenum, New York, 1983; A. F. Garito,
   Y. M. Chai, H. T. Man, and O. Zamani-Khamiri, p. 177 in Ref.
   8; G. M. Carter, M. K. Thakur, J. V. Hryniewicz, Y. J. Chen,
   and S. E. Meyler, p. 168 in Ref. 8.

10. L. Salem, *The Molecular Orbital Theory of Conjugated Systems*,
    Benjamin, New York, 1966. A thorough and accessible discussion
    of work up to 1965.

11. N. W. Ashcroft and N. D. Mermin, *Solid State Physics*, Holt-
    Saunders, New York, 1976.

12. G. Beni and P. Pincus, *Phys. Rev. B9*:2963 (1974); P. J.
    Strebel and Z. G. Soos, *J. Chem. Phys. 53*:4077 (1970); D. J.
    Klein and Z. G. Soos, *Mol. Phys. 20*:1013 (1971).

13. W. P. Su, J. R. Schrieffer, and A. J. Heeger, *Phys. Rev.
    Lett. 44*:1698 (1979); *Phys. Rev. B22*:2099 (1980); W. P. Su,
    in Ref. 1, Vol. 2, p. 757.

14. B. Roos and P. N. Skancke, *Acta Chem. Scand. 21*:233 (1967);
    K. Schulten, I. Ohmine, and M. Karplus, *J. Chem. Phys. 64*:
    4422 (1976); O. Chalvet, J. Hoarau, J. Joussot-Dubien, and
    J. C. Rayez, *J. Chim. Phys. 69*:630 (1972).

15. H. J. Longuet-Higgins and L. Salem, *Proc. Roy. Soc. London,
    Ser. A, 251*:172 (1959).

16. A. Streitwieser, *Molecular Orbital Theory for Organic Chemists*,
    Wiley, New York, 1961; C. A. Coulson and A. Streitwieser,
    Dictionary of π-Electron Calculations, Freeman, San Francisco,
    1965.

17. E. G. Wilson, *J. Phys. C: Solid State Phys., 8*:727 (1975);
    rather smaller transfer integrals are used than in Table 1.

18. J. A. Pople, *J. Chem. Phys., 37*:3009 (1962); K. Ruedenberg,
    J. Chem. Phys., 34:1986 (1961); p. 56–60 in Ref. 10.

19. J. Hubbard, *Proc. Roy. Soc. London, Ser. A, 276*:238 (1963); *277*:237 (1964); *281*:401 (1964).

20. K. Ohno, *Theor. Chim. Acta, 2*:219 (1964).

21. Z. G. Soos and D. J. Klein, *Treatise on Solid State Chemistry*, Vol. 3, N. B. Hannay, ed., Plenum, New York, 1976, p. 679.

22. O. Madelung, *Introduction to Solid-State Theory*, Springer-Verlag, Berlin, 1978, Ch. 8.

23. P. W. Anderson, *Phys. Rev. Lett., 34*:953 (1974); R. A. Street, and N. F. Mott, *Phys. Rev. Lett., 35*:1293 (1975); S. Robaszkiewicz, R. Micnas, and K. A. Chao, *Phys. Rev. B23*:1447 and references therein.

24. J. B. Sokoloff, *Phys. Rev. B2*, 779 (1970); G. Beni, T. Holstein, and P. Pincus, *Phys. Rev., B8*:312 (1973); D. J. Klein, *Phys. Rev., B8*:3452 (1973).

25. N. F. Mott, *Proc. Roy. Soc. London, Ser. A, 62*:416 (1949); *Rev. Mod. Phys., 40*:677 (1968); *Metal-Insulator Transitions*, Taylor and Francis, London, 1974.

26. J. H. Van Vleck, *Quantum Theory of Atoms, Molecules, and the Solid State*, P. O. Löwdin, ed., Academic Press, New York, 1966, p. 475.

27. P. W. Anderson, *Solid State Physics*, Vol. 14, F. Seitz and D. Turnbull, eds., Academic Press, New York, 1963, p. 99.

28. C. Domb and M. S. Green, eds. *Phase Transitions and Critical Phenomena*, Vol. 3, Academic Press, New York, 1974.

29. L. Pauling, *J. Chem. Phys., 1*:280 (1933); L. Pauling and G. W. Wheland, *J. Chem. Phys., 1*:362 (1933); G. W. Wheland, *Resonance in Organic Chemistry*, Wiley, New York, 1955.

30. S. Ramasesha and Z. G. Soos, *Inter. J. Quant. Chem., 25*: 1003 (1984). S. Mazumdar and Z. G. Soos, *Synth. Met., 1*:77 (1979).

31. Z. G. Soos and S. Ramasesha, *Phys. Rev. B29*:5410 (1984); *J. Chem. Phys., 80*:3278 (1984).

32. A. D. McLachlan, *Mol. Phys., 2*:276 (1959); J. Cizek, J. Paldus, and I. Hubac, *Intern. J. Quant. Chem., 8*:951 (1974).

33. O. J. Heilmann and E. H. Lieb, *Trans. N.Y. Acad. Sci., 33*: 116 (1971); S. R. Bondeson and Z. G. Soos, *J. Chem. Phys., 71*:380 (1979).

34. R. Pariser, *J. Chem. Phys., 24*:250 (1956); D. R. Herrick, *J. Chem. Phys., 74*:1239 (1981).

35. J. A. Pople and R. K. Nesbet, *J. Chem. Phys., 22*:571 (1954).

36. W. A. Goddard, II, *Phys. Rev., 157*:73, 81 (1967); R. C. Ladner and W. A. Goddard, II, *J. Chem. Phys., 51*:1073 (1969).

37. S. Kuwajima and Z. G. Soos, *J. Am. Chem. Soc., 109*:107 (1987).

38. M. A. Lee, S. Klemm, and S. Risser, *Condensed Matter Theories*, F. B. Malik, ed., Plenum, New York, 1986, p. 89; J. E. Hirsch, *Phys. Rev.*, *B32*:4403 (1985).

39. S. Masamune, F. A. Souto-Bachiller, T. Machiguchi, and J. E. Berlie, *J. Am. Chem. Soc.*, *100*:4889 (1978); W. T. Borden, E. R. Davidson, and P. Hart, *J. Am. Chem. Soc.*, *100*:388 (1978); L. J. Schaad, B. A. Hess, Jr., and C. Ewing, *J. Org. Chem.*, *47*:2904 (1982).

40. C. R. Fincher, Jr., C. E. Chen, A. J. Heeger, A. G. MacDiarmid, and J. B. Hastings, *Phys. Rev. Lett.*, *48*:100 (1982); C. S. Yannoni and T. C. Clarke, *Phys. Rev. Lett.*, *51*:1191 (1983). The X-ray and NMR data both indicate 2u $\sim$ 0.08 A for the difference between single and double bonds.

41. C. A. Coulson, *Proc. Roy. Soc. London, Ser. A. 164*:383 (1938); C. A. Coulson, *Proc. Roy. Soc. London, Ser. A, 169*: 413 (1939); see also Refs. 16 and 10.

42. R. E. Peierls, *Quantum Theory of Solids*, Clarendon, Oxford, 1955, p. 108.

43. E. H. Lieb and F. Y. Wu, *Phys. Rev. Lett.*, *20*:1445 (1968); M. Takahashi, *Prog. Theor. Phys.*, *42*:1098; *43*:1619 (1970).

44. A. A. Ovchinnikov, *Sov. Phys. JETP*, *30*:1100 (1970).

45. Z. G. Soos, S.Kuwajima, and J. E. Mihalick, *Phys. Rev.*, *B32*: 3124 (1985).

46. J. C. Bonner and H. W. J. Blöte, *Phys. Rev.*, *B25*:6959 (1982).

47. S. Mazumdar and S. N. Dixit, *Phys. Rev. Lett.*, *51*:292 (1983); S. N. Dixit and S. Mazumdar, *Phys. Rev.*, *B29*:1824 (1984).

48. N. A. Popov, *Zh. Strukt. Khim, 10*:533 (1969); A. Madhukar, *Solid State Commun.*, *15*:921 (1974); I. Egri, *Z. Phys.*, *B23*: 38 (1976); *Solid State Commun.*, *22*:281 (1977); M. H. Whangbo *J. Chem. Phys.*, *75*:2983 (1981).

49. S. Kivelson and D. Heim, *Phys. Rev.*, *B26*:4278 (1982); I. I. Ukranskii, *Zh. Eksp. Teor. Fiz, 75*:760 (1979) finds a slight enhancement; P. Horsch, *Phys. Rev.*, *B24*:7351 (1981) also finds enhancement.

50. J. E. Hirsch, *Phys. Rev. Lett.*, *51*:296 (1983);

51. D. Baeriswyl and K. Maki, *Phys. Rev.*, *B31*:6633 (1985).

52. G. W. Hayden and E. J. Mele, *Phys. Rev.*, *B32*:6527 (1985).

53. S. Kivelson, W. P. Su, J. R. Schrieffer, and A. J. Heeger, *Phys. Rev. Lett.*, *58*:1899 (1987); the choice of a δ-function potential and first-order perturbation theory have been questioned in Phys. Rev. Lett., 60:70 and 71 (1988).

54. C. A. Grob, *Acct. Chem. Res.*, *16*:426 (1983); H. C. Brown, *Acct. Chem. Res.*, *16*:432 (1983); A. Olah, G. K. Surya Prakash, and M. Saunders, *Acct. Chem. Res.*, *16*:440 (1983); C. Walling, *Acct. Chem. Res.*, *16*:448 (1983). A recent updating on the "Great Carbocation Problem."

55. J. A. Pople and S. H. Walmsley, *Mol. Phys.*, 5:15 (1962).
     See also p. 521–523 in Ref. 10 for an uncanny preview of SSH
     results.
56. A. A. Ovchinnikov, I. I. Ukranskii, and G. V. Kventsel, *Sov.
     Phys. Uspekhi, 15*:575 (1973).
57. D. W. Whitman and B. K. Carpenter, *J. Am. Chem. Soc., 104*:
     6473 (1982).
58. R. Jackiw and J. R. Schrieffer, *Nucl. Phys., B190* (FS3):253
     (1981).
59. H. Fukutome and M. Sasai, *Prog. Theor. Phys., 69*:373 (1983).
60. D. S. Boudreaux, R. R. Chance, J. L. Brédas, and R. Silbey,
     Phys. Rev., B28:6927 (1983).
61. T. Nakano and H. Fukuyama, *J. Phys. Soc. Japan, 49*:1679
     (1980); a continuum Heisenberg spin model is used and $\xi \sim 2.6$
     is found in this limit.
62. H. M. McConnell and D. B. Chesnut, *J. Chem. Phys., 27*:
     984, (1957); M. W. Hanna, A. D. McLachlan, H. H. Dearman,
     and H. M. McConnell, *J. Chem. Phys., 31*:361 (1962); A. Carring-
     ton, *Q. Rev., Chem. Soc., 17*:67 (1963).
63. A. Carrington and A. D. McLachlan, *Introduction to Magnetic
     Resonance*, Harper and Row, New York, 1967, p. 91.
64. H. Thomann and L. R. Dalton, ENDOR Studies of Polyacetylene,
     in Ref. 1, Vol. 2, p. 1157 and references therein. Both experi-
     mental and theoretical conflicts and/or misunderstandings remain
     to be resolved about the implications of negative spin densities.
65. H. Thomann, L. R. Dalton, Y. Tomkiewicz, N. J. Shiren, and
     T. C. Clarke, *Phys. Rev. Lett., 50*:553 (1983); L. R. Dalton,
     H. Thomann, A. Morrobel-Sosa, C. Chiu, M. E. Galvin, G. E.
     Wnek, Y. Tomkiewicz, N. S. Shiren, B. H. Robinson, and A. L.
     Kwiram, *J. Appl. Phys. Lett., 54*:5583 (1983).
66. A. J. Heeger and J. R. Schrieffer, *Solid State Commun., 48*:
     207 (1983).
67. K. J. Donovan, P. D. Freeman, and E. G. Wilson, *Mol. Cryst.
     Liq. Cryst., 118*:395 (1985); See also pp. 165 and 155 in Ref. 7;
     but D. Moses, M. Sinclair, and A. J. Heeger, *Phys. Rev. Lett.,
     58*:2710 (1987) find, on the basis of past transient photo-
     conductivity, that the mobility is orders of magnitude smaller.
68. J. L. Brédas, R. R. Chance, and R. Silbey, *Phys. Rev., B26*:
     5843 (1982). J. L. Brédas, B. Thémans, and J. M. André,
     Phys. Rev., B27:7827 (1983); *J. Chem. Phys., 78*:6137 (1983).
     See also Ref. 1, Vol. 2, and Refs. 5 and 6 for numerous and
     occasionally uncritical interpretation of optical, EPR, and trans-
     port data in terms of bipolarons.
69. H. Takayama, Y. R. Lin-Liu, and K. Maki, *Phys. Rev., B21*:
     2388 (1980).

70. S. A. Brazovskii, *JETP Lett.*, *28*:606 (1978); *Sov. Phys. JETP*, *51*:342 (1980).

71. J. A. Krumhansl, B. Horowitz, and A. J. Heeger, *Solid State Commun.*, *34*:945 (1980); B. Horowitz, *Solid State Commun.*, *34*: 61 (1980); *Phys. Rev. Lett.*, *46*:742 (1981).

72. D. K. Campbell, A. R. Bishop, and M. J. Rice, Ref. 1, Vol. 2, p. 937.

73. K. Fesser, A. R. Bishop, and D. K. Campbell, *Phys. Rev.*, *B27*:4804 (1983).

74. D. K. Campbell, D. Baeriswyl, and S. Mazumdar, *Synth. Met.*, *17*:197 (1987).

75. H. Sixl, Ref. 7, p. 41; *Mol. Cryst. Liq. Cryst.*, *134*:65 (1986); Ref. 2, p. 49.

76. M. J. Rice and E. J. Mele, *Phys. Rev. Lett.*, *49*:1455 (1982); D. K. Campbell, *Phys. Rev. Lett.*, *50*:865 (1983).

77. M. J. Rice, A. R. Bishop, and D. K. Campbell, *Phys. Rev. Lett.*, *51*:2136 (1983). M. Beer, *J. Chem. Phys.*, *25*:745 (1956); R. Eastmond, T. R. Johnson, and D. R. M. Walton, *Tetrahedron*, *28*:4601 (1972).

78. H. Köppel, L. S. Cederbaum, W. Dombke, and S. Shaik, *Angew. Chem. Int. Ed. Engl.*, *22*:210 (1983).

79. M. J. Rice and E. J. Mele, *Phys. Rev.*, *B25*:1339 (1982).

80. M. F. Granville, G. R. Holtom, and B. E. Kohler, *J. Chem. Phys.*, *72*:4671 (1980).

81. K. L. D'Amico, C. Manos, and R. L. Christensen, *J. Am. Chem. Soc.*, *102*:1777 (1980).

82. B. S. Hudson and B. E. Kohler, *Ann. Rev. Phys. Chem.*, *25*: 437 (1974); and references therein.

83. B. S. Hudson, B. E. Kohler, and K. Schulten, *Excited States*, Vol. 6, E. Lim, ed., Academic Press, New York, 1982.

84. S. Ramasesha and Z. G. Soos, *Chem. Phys.*, *91*:35 (1984).

85. J. Des Cloizeaux and J. J. Pearson, *Phys. Rev.*, *128*:2131 (1962).

86. Z. G. Soos and L. R. Ducasse, *J. Chem. Phys.*, *78*:4093 (1983).

87. B. R. Weinberger, C. B. Roxlo, S. Etemad, G. L. Baker, and J. Orenstein, *Phys. Rev. Lett.*, *53*:86 (1984).

88. H. Suzuki, *Electronic Absorption Spectra and Geometry of Organic Molecules*, Academic Press, New York, 1967.

89. S. Malhorta and M. Whiting, *J. Chem. Soc.*, *1960*:3812 (1960).

90. T. Sorenson, *J. Am. Chem. Soc.*, *87*:5075 (1965).

91. H. Kuhn, *J. Chem. Phys.*, *17*:1198 (1949); the wavelength λ of the $\pi \to \pi^*$ excitation goes as N in free electron theory. This works for dyes and ions, but not for polyenes where λ goes as $(a + b/N)^{-1}$. While the theoretical analysis is now dated, the distinction noted has been supported by many later

experiments. Free-electron (particle-in-box) ideas remain the first approximation for understanding conjugated systems. See also: J. R. Platt, *J. Chem. Phys.*, *17*:484 (1949); N. S. Bayliss, *J. Chem. Phys.*, *16*:287 (1948); W. J. Simpson, *J. Chem. Phys.*, *16*:1124 (1948); and Ref. 10.

92. J. Orenstein and G. L. Baker, *Phys. Rev. Lett.*, *49*:1042 (1982).

93. A. H. Zimmerman, R. Gygax, and J. I. Brauman, *J. Am. Chem. Soc.*, *100*:5595 (1978).

94. G. V. Uimin and S. V. Fomichev, *Sov. Phys. JETP*, *36*:1001 (1973).

95. A. J. Heeger, Ref. 1, Vol. 2, p. 729.

96. R. Silbey, Ref. 7, p. 93.

97. M. Kertész, *Adv. Quantum Chem.*, *15*:161 (1982).

98. M. H. Whangbo, R. Hoffmann, and R. B. Woodward, *Proc. Roy. Soc. London, Ser. A*, *366*:23 (1979); M. H. Whangbo, *Extended Linear Chain Compounds*, Vol. 2, J. S. Miller, ed., Plenum, New York, 1981.

99. J. L. Brédas, Ref. 1, Vol. 2, p. 859; J. L. Brédas, R. R. Chance, R. H. Baughman, and R. Silbey, *J. Chem. Phys.*, *76*:3673 (1982).

100. A. Karpfen, Ref. 7, p. 115.

101. J. Cizek and J. Paldus, *J. Chem. Phys.*, *47*:3976 (1967); J. Paldus and J. Cizek, *J. Chem. Phys.*, *52*:2919 (1970); J. Cizek, *Quantum Theory of Polymers*, J. M. André, J. Dehalle, and J. Ladik, eds., Reidel, Dordrecht, 1978, p. 103; M. Bénard and J. Paldus, *J. Chem. Phys.*, *72*:6546 (1980); J. Paldus, E. Chiu, and M. G. Grey, *Int. J. Quantum, Chem.*, *24*:395 (1983). See also Ref. 98 for summary.

102. S. Suhai, *Phys. Rev.*, *B27*:3506 (1983). Fully optimized HF geometry and Möller-Plesset perturbation theory for e-e contributions.

103. H. Fukutome and M. Sasai, *Prog. Theor. Phys.*, *67*:41 (1982); *69*:1 (1983); M. Sasai and H. Fukutome, *Prog. Theor. Phys.*, *70*:1471 (1983).

104. S. Suhai, *Phys. Rev.*, *B29*:4570 (1984).

105. S. Mazumdar and D. K. Campbell, *Phys. Rev. Lett.*, *55*: 2067 (1985); *Synth. Met.*, *13*:163 (1986).

106. R. Hoffmann, *J. Chem. Phys.*, *39*:1397 (1963); J. Ammeter, H. B. Burgi, J. C. Thibeault, and R. Hoffmann, *J. Am. Chem. Soc.*, *100*:3686 (1978).

107. M. J. S. Dewar and W. Thiel, *J. Am. Chem. Soc.*, *99*:4899 (1977); T. Clark, *A Handbook of Computational Chemistry*, Wiley, New York, 1985, Chaps 4 and 5 (current practice, references, and summary of semiempirical and ab initio methods); R. C. Bingham, M. J. S. Dewar, and D. H. Lo, *J. Am. Chem.*

Soc., *97*:1285 (1975). Excellent discussion of the MNDO/3
parameters and the relation between ab initio and semiempirical
methods.

108. J. A. Pople, D. P. Santry, and G. A. Segal, *J. Chem. Phys.*,
     43:S129 (1965); J. A. Pople and G. A. Segal, *J. Chem. Phys.*,
     43:S136 (1965); *44*:3289 (1966); J. A. Pople, D. L. Beveridge,
     and P. A. Dobosh, *J. Chem. Phys.*, *47*:2026 (1967); J. A.
     Pople and D. L. Beveridge, *Approximate Molecular Orbital
     Theory*, McGraw Hill, New York, 1970; W. J. Hehre, L. Radom,
     P. v. R. Schleyer, and J. A. Pople, *Ab Initio Molecular
     Orbital Theory*, Wiley, New York, 1986.

109. J. L. Brédas, R. Silbey, D. S. Boudreaux, and R. R. Chance,
     J. Am. Chem. Soc., *105*:6555 (1983); see also Ref. 99.

110. Y. S. Lee and M. Kertész, *Int. J. Quant. Chem.*, *Quant.
     Chem. Symp.*, *21*:163 (1987).

111. J. L. Brédas, A. J. Heeger, and F. Wudl, *J. Chem. Phys.*, *85*:
     4673 (1986); J. L. Brédas, *J. Chem. Phys.*, *82*:3808 (1982).

112. J. Terame, *Synth. Met.*, *19*:1004 (1987); A. Karpfen, *Physica
     Scripta*, *T1*:79 (1982).

113. J. L. Brédas, B. Thémans, and J. M. André, *Phys. Rev.*, *B27*,
     7827 (1983).

114. A. J. Epstein, Ref. 1, Vol. 2, p. 1041; K. L. Ngai and
     R. W. Rendell, Ref. 1, Vol. 2, p. 967.

115. G. Nicholas and P. Durand, *J. Chem. Phys.*, *70*:2020 (1979);
     72:453 (1980); J. M. André, L. A. Burke, J. Delhalle,
     G. Nicholas, and J. Durand, *J. Chem. Phys.*, *75*:255 (1981);
     J. L. Brédas, R. R. Chance, R. H. Baughman, R. Silbey,
     J. Chem. Phys., *76*:3673 (1982); see also Ref. 99.

116. R. R. Chance, D. S. Boudreaux, J. L. Brédas, and R. Silbey,
     Phys. Rev., *B27*:1440 (1983); Ref. 1, Vol. 2, p. 825.

117. C. T. White, F. W. Kutzler, and M. Cook, *Phys. Rev. Lett.*,
     56:252 (1986).

118. D. Vanderbilt and E. J. Mele, *Phys. Rev.*, *B22*:3939 (1980).

119. W. K. Wu and S. Kivelson, *Phys. Rev.*, *B33*:8546 (1986).

120. D. S. McClure, *J. Chem. Phys.*, *22*:1668 (1954); *24*:1 (1956);
     J. Wessel and D. S. McClure, *Mol. Cryst. Liq. Cryst.*, *58*:
     121 (1980).

121. P. R. Callis, T. W. Scott, and A. C. Albrecht, *J. Chem. Phys.*,
     78:16 (1983); T. W. Scott, P. R. Callis, and A. C. Albrecht,
     Chem. Phys. Lett., *93*:111 (1982); P. R. Callis, *Int. J. Quant.
     Chem: Quant. Chem. Symp.*, *18*:579 (1984).

122. H. B. Klevens and J. R. Platt, *J. Chem. Phys.*, *17*:470
     (1949).

123. C. A. Hutchison and B. W. Mangum, *J. Chem. Phys.*, *34*:
     908 (1961).

124. S. A. Boorstein and M. Gouterman, *J. Chem. Phys.*, *39*: 2443 (1963); A. G. Motten, E. R. Davidson, and A. Kwiram, *J. Chem. Phys.*, *75*:2603 (1981).

125. B. Dick and G. Olhlneicher, *Chem. Phys. Lett.*, *84*:471 (1981).

126. J. Tanaka and M. Tanaka, Ref. 1, Vol. 2, p. 1269; see also Ref. 4.

127. D. Bloor, *Photon, Electron, and Ion Probes of Polymer Structure and Properties*, D. W. Dwight, T. J. Fabish, and H. R. Thomas, eds., ACS Symposium 162, American Chem. Soc., Washington, D.C., 1983, p. 133; *Quantum Chemistry of Polymers—Solid State Aspects*, J. Ladik and J. M. André, eds., D. Reidel, Dordrecht, 1984, p. 191.

128. S. Roth and H. Bleier, *Synth. Met.*, *17*:503 (1987); see also Ref. 4a.

129. N. Basesau, Z. X. Liu, D. Moses, A. J. Heeger, H. Naarmann, and N. Theophilou, *Nature*, *327*:403 (1987); H. Naarmann, *Synth. Met.*, *17*:2233 (1987).

130. J. E. Hirsch and M. Grabowski, *Phys. Rev. Lett.*, *52*:1713 (1984); K. R. Subbaswamy and M. Grabowski, *Phys. Rev.*, *B24*:2168 (1981).

131. S. Kivelson and W. K. Wu, *Mol. Cryst. Liq. Cryst.*, *118*:9 (1985).

132. Z. G. Soos and S. Ramasesha, *Mol. Cryst. Liq. Cryst.*, *118*: 31 (1985).

133. A. S. Glick and G. W. Bryant, *Phys Rev.*, *B34*:943 (1986).

134. B. R. Weinberger, E. Ehrenfreund, A. Pron, A. J. Heeger, and A. G. MacDiarmid, *J. Chem. Phys.*, *72*:4749 (1980); Y. Tomkiewicz, T. D. Schultz, H. B. Brom, T. C. Clarke, and G. B. Street, *Phys. Rev. Lett.*, *43*:1532 (1979); K. Holczer, J. P. Boucher, F. Devreux, and M. Nechtschein, *Phys. Rev.*, *B23*:1051 (1981). Similar susceptibility and magnetic resonance data have been interpreted very differently.

135. L. R. Ducasse, T. E. Miller, and Z. G. Soos, *J. Chem. Phys.*, *76*:4094 (1982); Z. G. Soos, *Israel J. Chem.*, *23*:37 (1983).

136. J. R. Ackerman, B. E. Kohler, D. Huppert, and P. M. Rentzepis, *J. Chem. Phys.*, *77*:3967 (1982).

137. L. Rothberg, T. M. Jedhu, S. Etemad, and G. L. Baker, *Phys. Rev. Lett.*, *49*:3229 (1986).

138. D. Moses, M. Sinclair, and A. J. Heeger, *Synth. Met.*, *17*: 515 (1987); *Solid State Commun.*, *59*:343 (1986); see also Refs. 4a,c.

139. T. C. Clarke and J. C. Scott, Ref. 1, Vol. 2, p. 1127. A balanced and careful discussion of sometimes conflicting

interpretations of NMR data. See also H. Thomann, J. Haiyong, and G. L. Baker, *Phys. Rev. Lett.*, *59*:509 (1987) for evidence of similar soliton pinning in both *cis*- and *trans*-PA.

140. D. K. Campbell and A. R. Bishop, *Phys. Rev.*, *B24*:4859 (1981); *Nucl. Phys.*, *B200* (FS4):297 (1982); A. R. Bishop and D. K. Campbell, *Nonlinear Problems: Present and Future*, A. R. Bishop, D. K. Campbell, and B. Nicolaenko, eds., North Holland Mathematical Studies, Vol. 61, 1981, p. 195.

141. W. J. Feast, Ref. 1, Vol. 1, p. 1.

142. J. Orenstein, Ref. 1, Vol. 2, p. 1297.

143. T. Blum and H. Bässler, Ref. 8, p. 218; S. Etemad, G. L. Baker, J. Orenstein, and K. M. Lee, *Mol. Cryst. Liq. Cryst.*, *118*:389 (1985); see also Ref. 5.

144. T. D. Townsend and R. H. Friend, *Synth. Met.*, *19*:361 (1987).

145. R. R. Chance, J. L. Brédas, and R. Silbey, *Phys. Rev.*, *29*: 4491 (1984); Ref. 1, Vol. 2, p. 825.

146. S. T. Gentry and R. Kopelman, *J. Chem. Phys.*, *78*:373 (1980); *81*:3014, 3022 (1984) and references therein.

147. D. Emin, Ref. 1, Vol. 2, p. 936.

148. V. Dobrosavljevic and R. R. Stratt, *Phys. Rev.*, *B30*:2781 (1987).

149. J. P. Low, *Progr. Phys. Org. Chem.*, *6*:1 (1968); the cis-trans isomerization energy of butadiene is measured to be 0.11 ev and such values are typical for nominally single bonds. See also S. Sternhell, *Dynamic Nuclear Magnetic Resonance Spectroscopy*, L. M. Jackman and F. A. Cotton, eds., Academic Press, New York, 1975, p. 163.

150. H. Yamakawa, *Modern Theory of Polymer Solutions*, Harper and Row, New York, 1971.

151. Z. G. Soos and K. S. Schweizer, *Chem. Phys. Lett.*, *139*:196 (1987), Symposium of Electroactive Polymers, ACS-Denver Meeting (1987); (in Press). K. S. Schweizer and Z. G. Soos (unpublished).

152. H. Kuzamy, *Pure and A pl. Ch m.*, *57*:235 (1985); H. Kuzmany P, R. Surjan, and M. Kertész, *Solid State Commun.*, *48*:243 (1983); P. R. Surjan and H. Kuzmany, *Phys. Rev. B33*:2615 (1986).

153. E. Mulazzi, G. P. Brivio, and R. Tiziani, *Mol. Cryst. Liq. Cryst.*, *117*:343 (1985); E. Mulazzi, G. P. Brivio, S. Lefrant, and E. Faulques, *Mol. Cryst. Liq. Cryst.*, *117*:351 (1985); G. Brivio and E. Mulazzi, *Phys. Rev.*, *B30*:876 (1984).

154. G. N. Patel, R. R. Chance, and J. D. Witt, *J. Chem. Phys.*, *70*:4387 (1979); see also Ref. 7.

155. L. A. Harrah and J. M. Zeigler, *J. Poly. Sci. Poly. Lett.*, *23*:209 (1985); *Macromolecules*, *20*:601 (1987); R. D. Miller, D. Hofer, J. Rabolt, and G. N. Ficknes, *J. Am. Chem. Soc.*,

107:2175 (1985); P. Trefonas, J. R. Dameswood, and R. D. Miller, *Organometallics*, 4:1318 (1985).

156. R. H. Friend, H. E. Schaffer, A. J. Heeger, and D. C. Bott, *J. Phys. C. Solid State Phys.*, 20:6013 (1987).

157. A. Ozkabak, L. Goodman, S. Thakur, and K. Krogh-Jespersen, *J. Chem. Phys.*, 83:6047 (1985) and references therein.

158. E. J. Mele and M. J. Rice, *Phys. Rev. Lett.*, 46:926 (1980); E. J. Mele, Ref. 1, Vol. 2, p. 795 (1986).

159. M. Takahashi and J. Paldus, *Can. J. Phys.* 62:1226 (1984).

160. M. J. Rice, N. O. Lipari, and S. Strässler, *Phys. Rev. Lett.*, 21:1359 (1977); M. J. Rice, L. Pietronero, and P. Brüesch, *Solid State Commun.*, 21:757 (1977); R. Bozio and C. Pecile, *Physics and Chemistry of Low-Dimensional Solids*, NATO ASI Series C56, L. Alcacer, ed., D. Reidel, Dordrecht, 1980, p. 165; A. Girlando, A. Painelli, and C. Pecile, *Mol. Cryst. Liq. Cryst.*, 120:17 (1985).

161. B. Horowitz, Z. Vardeny, and O. Brafman, *Synth. Met.*, 9: 215 (1984).

162. B. Horowitz, *Solid State Commun.*, 41:729 (1982); E. Ehrenfreund, Z. Vardeny, O. Brafman, and B. Horowitz, *Mol. Cryst. Liq. Cryst.*, 117:367 (1986); Z. Vardeny, O. Brafman, E. Ehrenfreund, and B. Horowitz, *Phys. Rev. Lett.*, 51:2326 (1983); 54:75 (1985).

163. D. N. Batchelder, Ref. 7, p. 187.

164. E. Ehrenfreund, Z. Vardeny, O. Brafman, B. Horowitz, J. Tanaka, and M. Tanaka, *Synth. Met.*, 13:263 (1987).

165. Z. G. Soos, and G. W. Hayden (to be published); N. A. Fisher, Senior Thesis, Princeton University (1987), unpublished.

166. G. Wegner, *Z. Naturforschg.*, 24b:824 (1969); G. Wegner, *Molecular Metals*, W. E. Hatfield, ed., Plenum, New York, 1979, p. 209 and references therein. See also Ref. 7.

167. V. Enkelmann, Ref. 2, p. 137.

168. F. Ebisawa, T. Kurihara, and H. Tabei, *Synth. Met.*, 18:431 (1987); U. Seiferheld and H. Bässler, *Solid State Commun.*, 47: 391 (1983); H. Nakanishi, F. Mizutani, M. Kato, and H. Hasumi, *J. Poly. Sci. Poly. Lett.*
D. J. Sandman, and M. A. Newkirk, *Appl. Phys. Lett.*, 46: 100 (1985); N. Ferrer-Anglada, D. Bloor, I. F. Chalmers, I. G. Hunt, and R. D. Hercliffe, *J. Mater. Sci. Lett.*, 4:83 (1985).

169. Z. G. Soos, S. Mazumdar, and S. Kuwajima, Ref. 8, p. 190.

170. B. Reimer and H. Bässler, *Chem. Phys. Lett.*, 55:315 (1978); R. R. Chance, M. L. Shand, C. Hogg, and R. Silbey, *Phys. Rev.*, B22:3540 (1980); these two-photon results place 2^1A_g above E_g; Y. Tokura, Y. Oowaki, T. Koda, and R. H. Baughman, *Chem. Phys.*, 88:437 (1984); electro-reflectance is

interpreted as 2^1A_g above E_g; P. A. Chollet, F. Kajzar, and J. Messier, *Synth. Met.*, *18*:459 (1987); $\chi^{(3)}$ is interpreted as 2^1A_g below E_g.

171. D. B. Fitchen, *Synth. Met.*, *9*:341 (1984) suggest $E(2^1A_g)$ $\sim$ 1.2 ev; P. Tavan and K. Schulten, *J. Chem. Phys.*, *70*: 5407 (1979) place it $\sim$0.7 ev below E_g; F. Kajzar, S. Etemad, G. L. Baker, and J. Messier, *Synth. Met.*, *17*:563 (1987) interpret $\chi^{(3)}$ to put $E(2^1A_g)$ at 1.9 ev, essentially at E_g.

172. K. Lochner, B. Reimer, and H. Bässler, *Phys. Sat. Sol.*, *76b*: 533 (1976); V. Enkelmann and G. Wegner, *Makromol. Chem.*, *178*:635 (1977).

173. S. Etemad, T. Mitani, M. Ozaki, T. C. Chung, A. J. Heeger, and A. G. MacDiarmid, *Solid State Commun.*, *40*:75 (1981); T. Tani, P. M. Grant, W. D. Gill, G. B. Street, and T. C. Clarke, *Solid State Commun.*, *33*:499 (1980).

174. H. Sixl, Ref. 8, p. 12.

175. L. Robins, J. Orenstein, and R. Superfine, *Phys. Rev. Lett.*, *56*:1850 (1986).

176. G. J. Exarhos, W. M. Risen, Jr., and R. H. Baughman, *J. Am. Chem. Soc.*, *98*:481 (1976).

177. K. K. Georgieff, W. T. Cave, and K. G. Blaikie, *J. Am. Chem. Soc. 76*:5494 (1954); *UV Atlas of Organic Compounds*, Vol. III, Plenum, New York, 1967.

178. U. Dinur and M. Karplus, *Chem. Phys. Lett.*, *88*:171 (1982).

179. Y. Tokura, T. Mitani, and T. Koda, *Chem. Phys. Lett.*, *75*: 324 (1980); Y. Oowaki, Y. Kaneko, T. Koda, and T. Mitani, *J. Phys. Soc. Japan*, *53*:4054 (1984).

180. Z. G. Soos, S. Mazumdar, and S. Kuwajima, *Physica*, *143B*: 538 (1986).

181. G. W. Hayden and E. J. Mele, *Phys. Rev.*, *B34*:5484 (1986).

182. P. Tavan and K. Schulten, Phys. Rev. B. 36:4337 (1987).

183. M. Kertész and R. Hoffmann, *Solid State Commun.*, *47*:97 (1983); S. Kilvelson and O. L. Chapman, *Phys. Rev.*, *B28*: 7236 (1983); S. Aono, K. Nishikawa, M. Kimura, and H. Kawabe, *Synth. Met.*, *17*:167 (1987); M. Kimura, H. Kawabe, K. Nishikawa, and S. Aono, *J. Chem. Phys.*, *85*:3097 (1986).

184. D. J. Klein, T. G. Schmalz, G. E. Hite, A. Metropoulos, and W. A. Seitz, *Chem. Phys., Lett.*, *120*:367 (1985).

Index